Reactivity and Structure
Concepts in Organic Chemistry

Volume 26

Editors:

Klaus Hafner   Jean-Marie Lehn
Charles W. Rees   P. von Rague Schleyer
Barry M. Trost   Rudolf Zahradník

Leo A. Paquette · Annette M. Doherty

Polyquinane Chemistry

Syntheses and Reactions

With 140 Schemes

Springer-Verlag
Berlin Heidelberg New York
London Paris Tokyo

Professor Leo A. Paquette
Dr. Annette M. Doherty

Evans Chemical Laboratories
The Ohio State University
Columbus, Ohio 43210/USA

CHEM

sep/ae

ISBN 3-540-17703-5 Springer-Verlag Berlin Heidelberg New York Tokyo
ISBN 0-387-17703-5 Springer-Verlag New York Heidelberg Berlin Tokyo

Library of Congress Cataloging – in – Publication Data
Paquette, Leo A. Polyquinane chemistry. (Reactivity and structure : concepts in organic
chemistry ; v. 26) Bibliography: p. Includes index.
1. Quinanes. I. Doherty, Annette M. (Annette Marian), 1961–. II. Title. III. Series:
Reactivity and structure ; v. 26.
QD335–.P33 1987   547′.593   87–9471
ISBN 0–387–17703–5 (U.S.)

© by Springer-Verlag Berlin Heidelberg 1987
Printed in Germany

Typesetting: Friedrich Pustet, Regensburg. Printing: Mercedes-Druck, Berlin.
Bookbinding: Lüderitz & Bauer, Berlin.
2152/3020-543210

List of Editors

Professor Dr. Klaus Hafner
Institut für Organische Chemie der TH Darmstadt
Petersenstr. 15, D-6100 Darmstadt

Professor Dr. Jean-Marie Lehn
Institut de Chimie, Université de Strasbourg
1, rue Blaise Pascal, B.P. 296/R8, F-67008 Strasbourg-Cedex

Professor Dr. Charles W. Rees, F. R. S. Hofmann
Professor of Organic Chemistry, Department of Chemistry
Imperial College of Science and Technology
South Kensington, London SW7 2AY, England

Professor Dr. Paul v. Ragué Schleyer
Lehrstuhl für Organische Chemie der Universität Erlangen-Nürnberg
Henkestr. 42, D-8520 Erlangen

Professor Barry M. Trost
Department of Chemistry, The University of Wisconsin
1101 University Avenue, Madison, Wisconsin 53 706, U.S.A.

Professor Dr. Rudolf Zahradník
Tschechoslowakische Akademie der Wissenschaften
J.-Heyrovský-Institut für Physikal. Chemie und Elektrochemie
Máchova 7, 121 38 Praha 2, C.S.S.R.

Table of Contents

I Introduction

New methods for the construction of condensed five-membered ring systems continue to be developed at an accelerated pace. The challenges underlying this tremendous current upsurge of interest arise from several directions. One stems from the desire to elucidate and resolve those special problems associated with the incorporation of added strain not present when six-membered rings are mutually fused. The many structurally interesting polyquinane natural products isolated and characterized in recent years have provided a particularly delightful forum for application of various new synthetic protocols, many of which must equally well accommodate the particular stereochemical demands of each individual target. Synthetic elaboration of a marvellous array of new unnatural molecules also holds continued fascination.

In the past, we have attempted to keep others abreast of developments in this rapidly burgeoning area by authoring a pair of comprehensive reviews in Topics in Current Chemistry that appeared in 1979 [1] and 1984 [2]. During this period, others have also surveyed the developments in cyclopentannulation [3] and the cyclopentanoid field in general [4]. In the last couple of years, the pace at which new synthetic facets have been reported has become more frenetic than ever before. Accordingly, a suitable updating of the exciting newer findings was deemed appropriate and the present overview, which extends approximately to mid — 1986, was written. Once again, our hope is that compilations of this type will serve to stimulate imaginative new scientific ventures that will propel the field forward to still greater maturity.

Acknowledgments

In writing a book about recent developments in the chemistry of polyquinane systems, we are conscious of the very considerable contributions to this subject by our colleagues throughout the world. In many ways, this book stands as a testimony to their research accomplishments. In an undertaking of this magnitude, especially in view of the large number of illustrations, some errors will be inevitable. We hope that these do not otherwise detract from the intended impact of the overall theme.

The enormously ardous tasks of drawing all of the chemical structures and of typing the entire manuscript were the master work of Mrs. Donna Rothe, to whom we publicly express our deepest gratitude and appreciation.

II New Synthetic Developments

A Annulation Reactions

1 Acid- and Base-Promoted Cyclizations

Lee and Richardson have disclosed that it is entirely feasible to cyclopentannulate silyl enol ethers in a one-pot two-step process. For example, condensation of 1 with the readily available stannane 2 in the presence of trimethylsilyl triflate leads to an adduct that cyclizes to 3 on subsequent exposure to titanium tetrachloride [5]. This new strategy is notable in that a doubly functionalized product results. 1,2-Bis(trimethylsilyloxy)cyclopentene (4) condenses with chloride 5 in the presence of anhydrous zinc chloride to give a 1:1 mixture of 6 and 7 [6]. The first relevant point is that alkylation proceeds without annihilation of the allylsilane moiety. Consequently, this functionality is available for subsequent ring closure as illustrated by the independent conversion of 6 and 7 to 8 and 9, respectively, when treated with ethylaluminum dichloride.

Jager and his coworkers have developed an intriguing and highly serviceable iterative cyclopentane annulation of α,β-unsaturated ketones [7]. The process,

which is illustrated in Scheme I, features as its key steps conversion to an allylic alcohol by hydride reduction or Grignard addition, Claisen rearrangement via either the ortho ester or ketene acetal modification, and polyphosphoric acid-induced cyclization of the derived carboxylic acid.

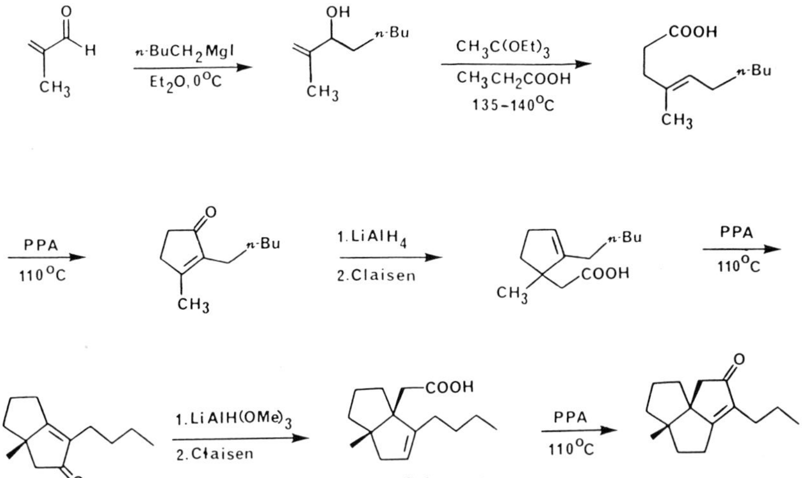

Scheme I

Use of the intramolecular Wadsworth-Emmons reaction for fusing a cyclopentenone ring can sometimes prove quite troublesome. Aristoff has demonstrated that avoidance of the use of strong base and substitution of one equivalent of potassium carbonate and two equivalents of 18-crown-6 in warm toluene may alleviate this problem [8]. The 70% yield realized in the formation of 10 is apparently indicative of the efficiency of this process. It is widely recognized that the presence of an angular substituent causes aldolization to proceed less arduously. The cyclization of 11 constitutes an additional example [9].

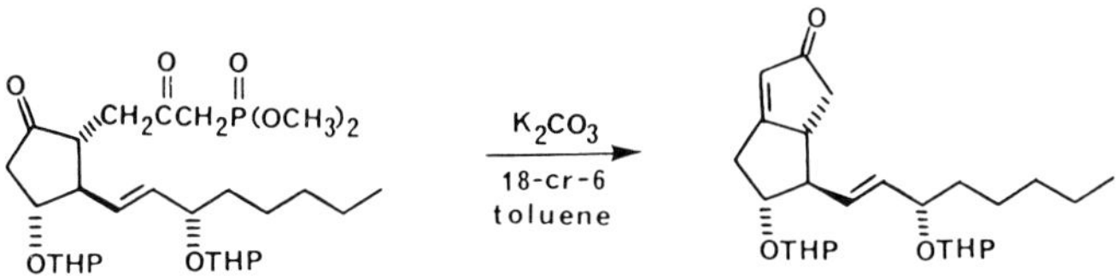

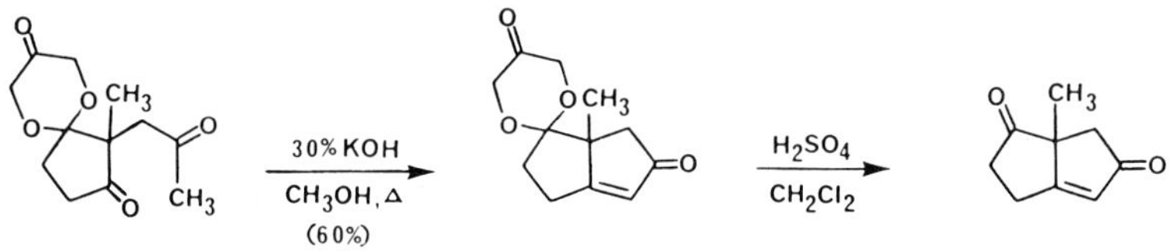

The elusive ketone 12 has been generated by cyclization of a polymer-bound precursor and liberation under strongly basic conditions [10]. Through use of the three-phase test, 12 was shown to be capable of acting as a diene but not as a dienophile in select pericyclic reactions.

3

12

It now appears, on the basis of several reports, that cycloalkylation of a doubly activated cyclopentane can deliver diquinanes efficiently. Certainly the dianions of several 3-isobutoxycyclopent-2-en-1-ones lend themselves readily to *cis*-bicyclo-[3.3.0]octenone construction [11]. Dimethyl 1,2-cyclopentanedicarboxylate is equally compatible, condensation of its dienolate with ethyl 3-bromopropionate leading chiefly to 13 [12]. The dilithiated bisoxazoline 14 is particularly serviceable because it allows for stepwise fusion of two differently functionalized five-membered rings. The synthesis of 15 exemplifies the latent promise of this intermediate [13].

13

14

15

The pair of imides 16 and 17 offer sufficiently different degrees of endo steric shielding to exhibit widely divergent annulation stereochemistry [14]. 1,1-Dilithioallyl phenyl sulfone has been developed as an efficient reagent for cyclopentane ring construction; the formation of 18 in 82% yield is an indicator of its propensity for geminal cycloalkylation [15]. 1,1-Bis(benzenesulfonyl)cyclo-

propane has been shown by Trost and his coworkers to be capable of nucleophilic ring opening [16]. In the case of keto ester 19, alkylation can be followed successfully by *in situ* methylation to give 20. Reduction of this intermediate with lithium phenanthrenide in tetrahydrofuran at −78 °C resulted in cyclization and formation of bifunctional diquinane 21.

Nitro compounds are being utilized with increasing frequency in five-ring annulation reactions. For example, the Lewis acid-promoted condensation of silyl enol ethers with aliphatic nitro olefins affords 1,4-diketones in good yield. The latter intermediates can be efficiently transformed into fused cyclopentenones when an angular substituent is present as shown by the synthesis of 22 [17]. Reaction of 2-methyl-1,3-cyclopentanedione with 2-nitropropene in the presence of catalytic amounts of tri-*n*-butylphosphine leads quantitatively to the Michael adduct 23. Once monoketalization is accomplished, the nitro group can

be transposed into a carbonyl center under alkaline conditions and cyclization to 24 subsequently realized [18]. Bicyclo[3.3.0]octane-2,8-dione (25) is most conveniently prepared by addition of ethyl 3-nitropropionate to 2-cyclopentenone, hydrogenation of the resulting acrylate, and intramolecular Claisen condensation [19]. Angularly substituted derivatives of 25 have been found responsive to asymmetric induction by enantiotopic differentiation in retro-Claisen reactions involving optically active bases [20].

The utility of phosphorus reagents for achieving cyclopentannulation continues to make its appearance. Hewson and McPherson have shown that vinyl phosphonium salts 26 a and 26 b offer considerable potential for preparing highly functionalized diquinanes such as 27 [21]. An interesting new process developed by Bestmann involves reaction of cyclopentanone acetic acids exemplified by 28 and 29 with triphenyl[(phenylimino)ethenylidene]phosphorane (30). Subsequent heating of the resulting ylides delivers the target diquinanes efficiently [22]. Another ingenious cyclopentannulation scheme begins with the bicyclic intermediates 31, derived from tricthylsilyloxycyclopentenes and ethyl diazoacetate.

Treatment of 31 with vinyl phosphonium salts 32 a or 32 b in the presence of potassium fluoride and a catalytic amount of 18-crown-6 in refluxing acetonitrile delivered 33. When Y in 33 was methylthio, further hydrolysis to the corresponding bicyclo- and tricyclooctanones 34 and 35, respectively, was possible [23].

In a different context, cyclization of keto sulfoxides under the conditions of the Pummerer reaction has been shown to proceed effectively as illustrated in the synthesis of diquinane 36 [24].

2 Metal-Promoted Ring Closures

The modified Barbier reaction can be adapted to the stereoselective construction of di- and triquinanes. The cyclization expectedly shows a strong preference for formation of the cis alcohols as illustrated below [25, 26]. A related intramolecular ring closure has been reported for allylic halides such as 37 when these substances are exposed to the combined action of metallic tin and aluminum in an aqueous organic solvent system [27]. Corey and Pyne have been successful in reversing the process and realizing five-ring annulation by reduction of a carbonyl group followed by internal addition of the resulting free radical to an adjacent π bond [28]. Thus, zinc metal and trimethylchlorosilane act on 38–43 in the presence of lutidine to give diverse diquinanes. The process appears to tolerate numerous functional groups well, although it is nonstereoselective for the pendant group at C-2.

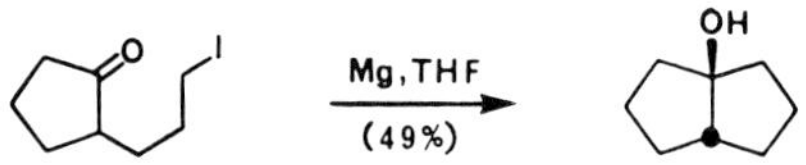

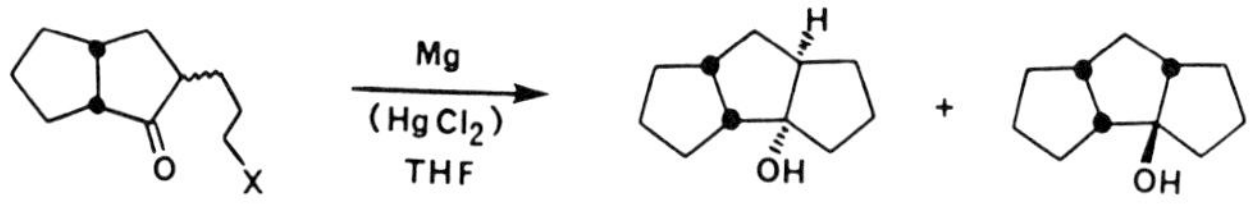

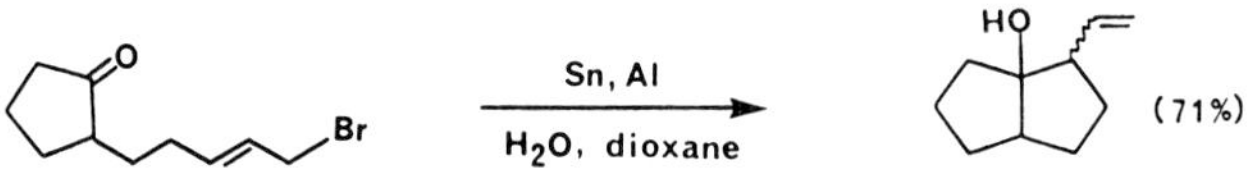

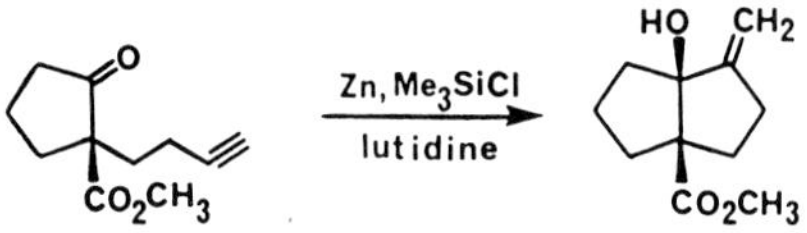

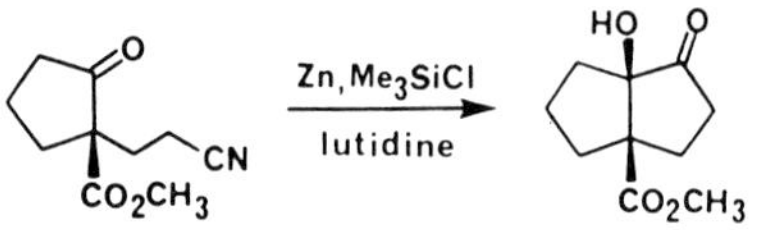

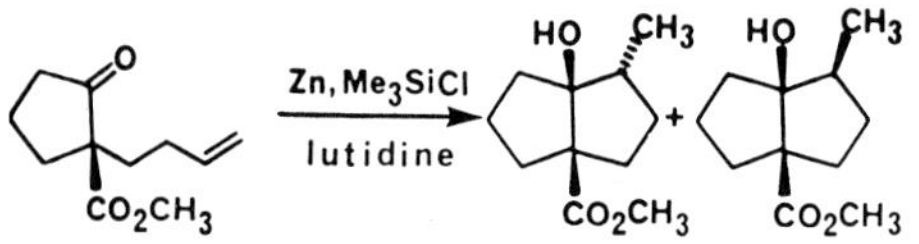

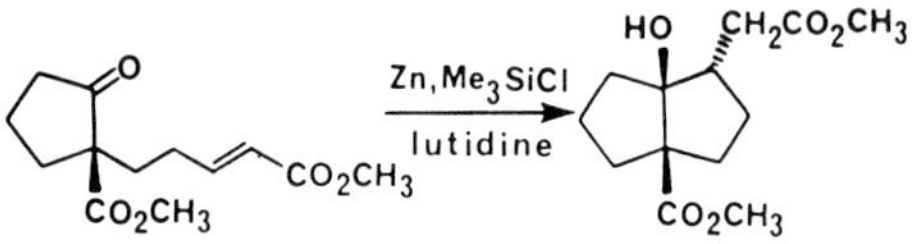

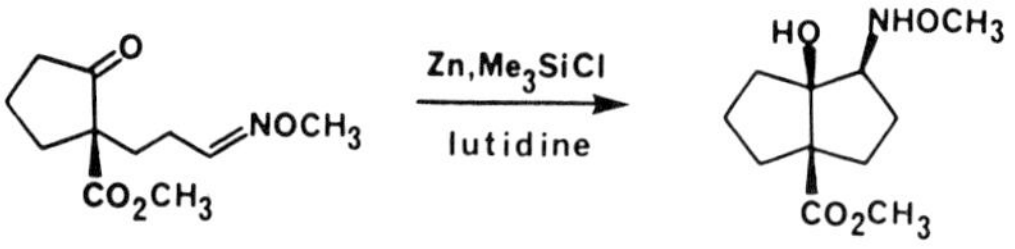

43

Two groups have uncovered that the latter cyclization can also be realized electrochemically, at least when two types of acceptor sidechains are present. In the first instance, electrolysis of terminal allenic ketones of type 44 resulted in smooth cyclization to give 45. Comparable treatment of the homologue (46) led only to 47, thereby demonstrating that closure in the «exo-mode» operates to the exclusion of six-ring formation [29]. Electroreduction of systems having unsaturated esters tethered to the cyclopentanone rings as in 48 likewise results in intramolecular cyclization [30].

44   **45**   (42%)

46   **47**   (23%)

(79%)

48   (cis/trans = 11.4)

Reduction of 49 with sodium amalgam in methanol buffered with disodium hydrogen phosphate leads to a mixture of four volatile hydrocarbons, one of which (17%) is the diquinane 50 [31]. Treatment of 51 with *n*-butyllithium in

49   **50**

51   **52**   **53**

tetrahydrofuran at $-78\,°C$ gives the dianion, exposure of which to cupric chloride results in oxidative coupling to give 52. Subsequent two-fold deprotonation of 52 followed by iodine oxidation delivers the novel cyclopenta[*a*]pentalene 53 as dark red needles [32]. The 12π-electron anion of 53 is resonance-stabilized.

Carbolactonization of keto acid 54 with manganese(III) acetate produces tricyclic keto lactone 55, which can be further modified chemically to afford tetracyclic dilactone 56 [33].

3 Free Radical Processes

Intramolecular homolytic addition reactions involving two consecutive ring closures sometimes give only one or two major products and are of great potential utility for synthesis (see later). Beckwith and coworkers have examined the simplest test case for linear triquinane construction via three consecutive cyclizations. In line with force-field calculations, which did not provide a firm basis for predicting the stereochemical outcome of this process [34], the heating of bromide 57 in benzene with tributylgermane and a trace of AIBN led to an eight-component hydrocarbon mixture. However, the two *cis,anti,cis*-tricyclo[6.3.0.0^{2,6}]undecanes 58 and 59 did predominate (27%) [35].

Winkler and Sridar have assessed the level of stereochemical control attainable in transannular radical cyclizations also leading to linearly fused cyclopentanoids [36]. Irradiation of 60 in the presence of tri-*n*-butyltin hydride led to formation of four hydrocarbons which reflected a kinetic preference for trans-fused products (73:27) as a result of conformational control. As expected from thermodynamics, trans radical 61 does not continue to react transannularly. On the other hand, cis radical 62 partitions between reduction and formation of 63 and 64 in a 1:2 ratio. Treatment of 65 with di-*tert*-butyl peroxide at $150\,°C$

60   (n-Bu)₃SnH (AIBN) hυ   61   +   62

63   +   64

65   (t-BuO)₂ 150 °C   66   +   67

68   (t-BuO)₂ 150 °C   69

delivered 66 and 67 in a 2.8:1 ratio. The exclusive formation of cis-anti-cis triquinanes is interesting. In comparable fashion, 68 leads in 45% yield to 69.

The sequential treatment of α,β-unsaturated ketones with (phenylseleno)alanes or titanates followed by an α,γ-unsaturated aldehyde gives rise efficiently to β-(phenylseleno)ketols exemplified by 70 and 72. Reductive cyclization of these entities in the presence of tri-n-butyltin hydride affords the corresponding condensed diquinanes 71 and 73, respectively, in high yields [37]. The radical intermediates generated by appropriate deoxygenation of alcohols have been found capable of intramolecular ring closure when suitably located triple bonds (C≡C or C≡N) are present. With 74, the yield is lower than those encountered with larger rings [38]. In a related development, β-acetylenic radicals react with electron-deficient olefins to produce diquinanes such as 75 by a process involving conjugate addition followed by 5-exo-digonal closure [39].

70   1. PhSeAl(CH₃)₂   2. CHO   (n-Bu)₃SnH (AIBN) toluene, Δ   71

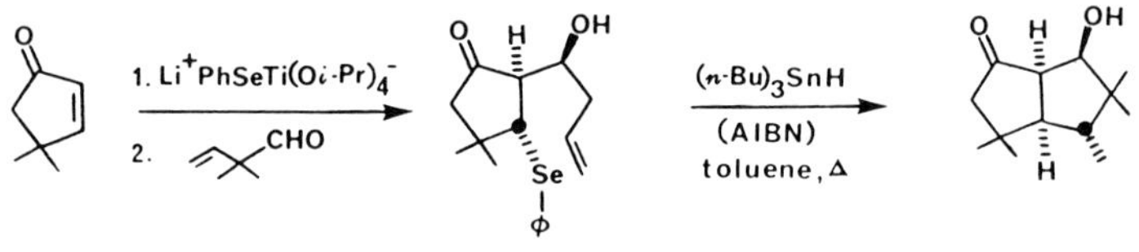

72　　　　　　　　73

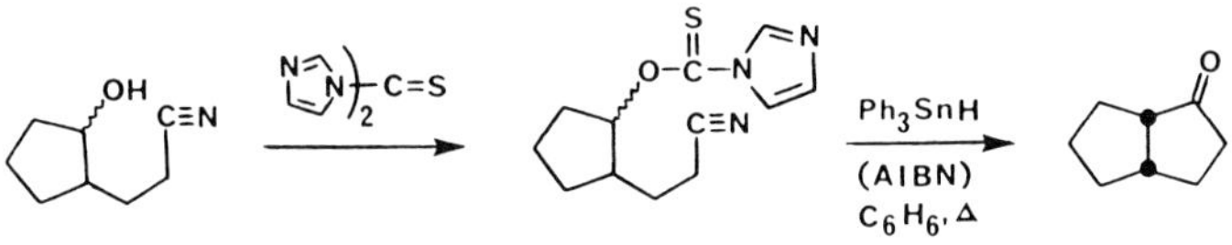

74　　　　　　　(15%)

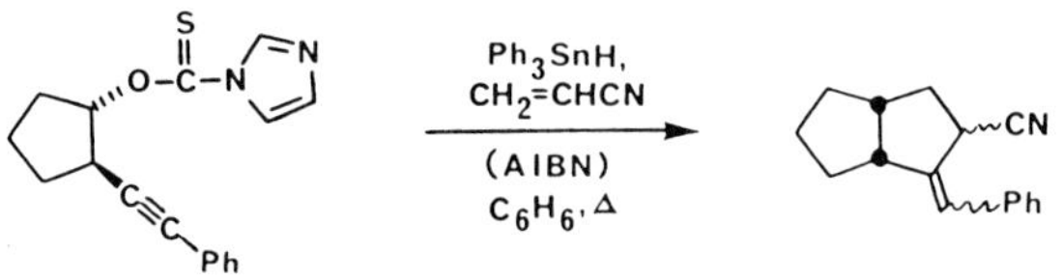

75

δ,ε-Unsaturated ketones are efficiently photoreduced in HPMA or tertiary amines in a reaction that results only in cyclization with formation of a cyclopentanol ring [40].

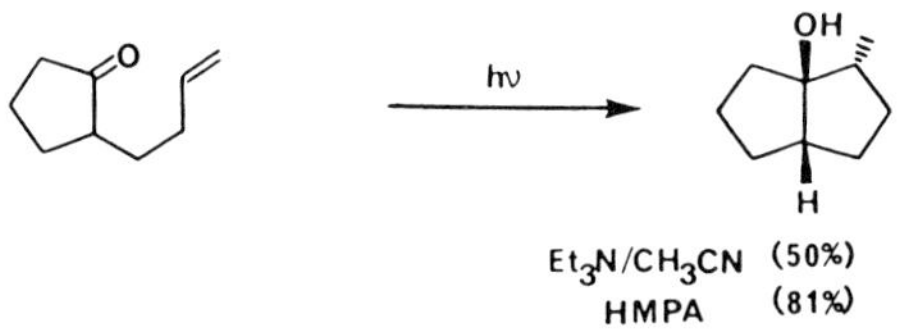

Et$_3$N/CH$_3$CN (50%)
HMPA (81%)

4 [3 + 2] Cycloaddition Methodology

(2-(Acetoxymethyl)-3-allyl)trimethylsilane (76) in the presence of a palladium(0) reagent serves as an equivalent of trimethylenemethane in cycloadditions to electron-deficient olefins such as α,β-unsaturated ketones, esters, nitriles, sul-

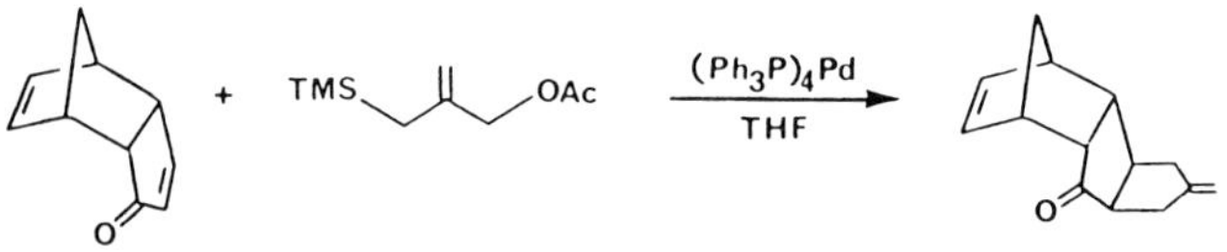

76　　　　　　(52%)

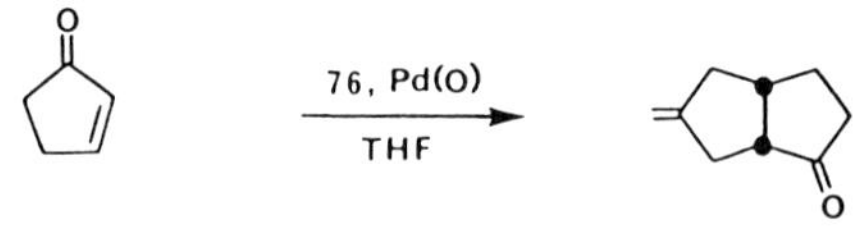

(56%)

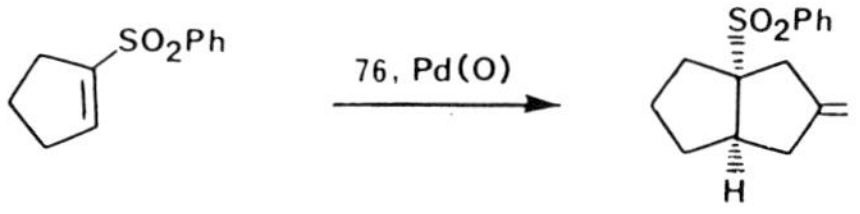

(58%)

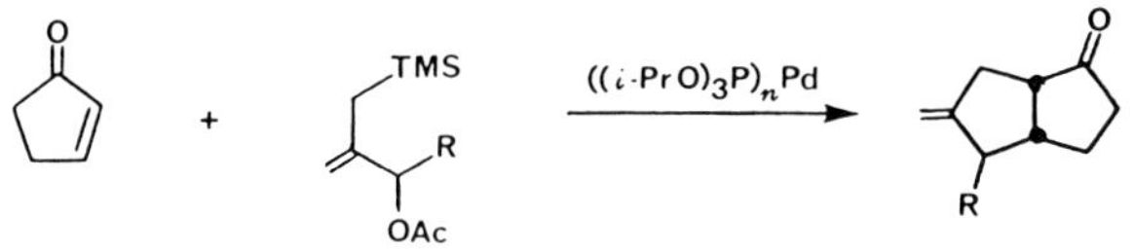

77a, R = CH=CH₂
 b, R = C₆H₅

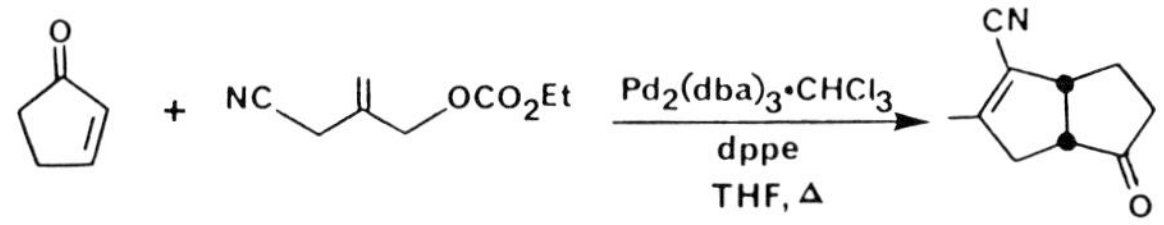

78   (71%)

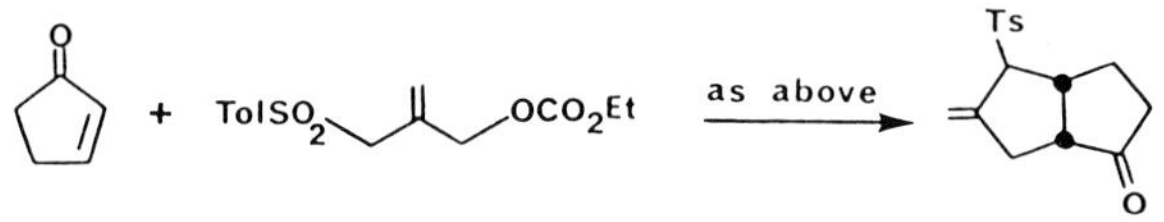

79   (65%)

fones, and lactones [41]. The vinyl (77 a) [42], phenyl (77 b) [42], cyano (78) [43], and p-toluenesulfonyl analogues (79) [43] of 76 have also been shown to add regioselectively to give highly functionalized diquinanes in good yields.

Palladium(0)-catalyzed reaction of vinylcyclopropanes carrying two electron-withdrawing groups (e. g., 80) with cyclic enones results in efficient cyclo-pentannulation.

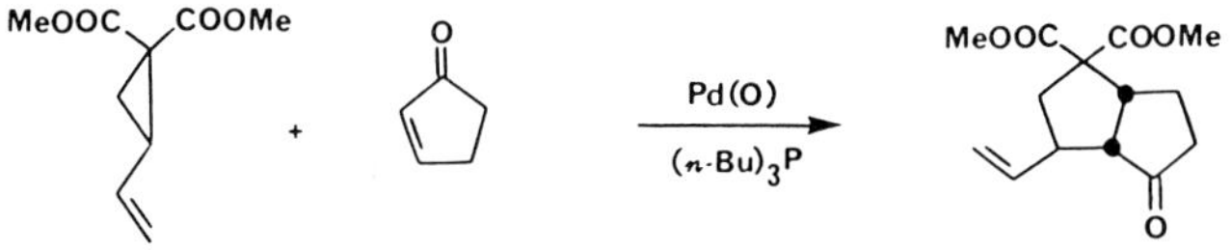

80   (87%)

A regiospecific [3 + 2] annulation approach to highly substituted diquinanes has been developed by Danheiser and his students [44]. The process involves reaction of (trimethylsilyl)allenes with electron-deficient alkenes and alkynes in the presence of titanium tetrachloride.

(48%)

13

Saito and coworkers have observed that 2-cyanochromone (81) gives rise upon irradiation in the presence of olefins to mixtures of [3 + 2] and [2 + 2] cycloadducts [45].

5 Weiss-Cook Condensations

Condensation of dimethyl 3-ketoglutarate with the C_{22} α-diketone 82 and subsequent acidic hydrolysis and decarboxylation has afforded 83, probably the largest propellane known [46]. The key structural intermediate in these reactions has been identified [47], and the mechanism of formation of 84 and 85 from glyoxal has been resolved [48]. Lastly, steric factors present in the 1,2-diketone have been found to play a major role in the overall success of the reaction [49].

6 Intramolecular Diels-Alder Processes

The Sternbach group has examined the feasibility of an intramolecular Diels-Alder reaction involving cyclopentadienes of type 86 followed by cleavage of the double bond in adduct 87 as a route to bicyclo[3.3.0]octanes. Three quaternary centers could be formed during the [4+2] cycloaddition depending on the substitution at the dienophile, and the stereochemistry of the four chiral centers shown in 88 would then become well defined [50]. Heating 89 a at 110 °C for 4 h and direct hydrolysis yielded isomerically pure 91 in 26% overall yield. In the case of 89 b, reaction at 160 °C for 5 h afforded 90 b more efficiently (60%).

An efficient reaction occurs when 92 is heated at 160 °C for 88 h. The cyclo-addition is much faster when the quaternary carbon on the bridging chain is positioned as in 93 (4 h for completion). The substitution pattern in 94 accelerates the process still more (2 h at 160 °); after hydrolysis, ketone 95 was isolated in 74% overall yield [50].

In a variant of this general concept, Alward and Fallis heated ketone 96 at 180 °C for 7 days. Cycloaddition proceeded via 97 to deliver 98 in 53% yield [51].

Heating acetate 99 in mesitylene at 220 °C for 50 h (sealed tube) assembles the tricyclic ketones 100 and 101 in a 14.6:1 ratio after alkaline hydrolysis [52]. Following conventional conversion of the major product to 102, contraction of the cyclohexanone ring can be implemented to arrive at 103, a molecule possessing the basic cedrane carbocyclic framework.

7 Cyclization of Diazo Compounds

Whereas copper(I) triflate has been touted as the transition metal catalyst of choice for cyclization of simple diazo ketones related to 104 [53], copper powder in refluxing cyclohexane affords the best results when α-diazo-β-ketophosphonates (e. g., 105) are involved [54]. Cyclization of the mixture of diazo ketones

106   107   108

py·HCl, Δ

109

represented by 106 with cupric sulfate in a benzene-cyclohexane solvent system provides 107 (50%) and 108 (38%). Heating of this mixture with pyridinium hydrochloride opens an interesting access route to diquinane 109 [55].

The aziridinyl imine generated by reaction of 110 with 1-amino-*trans*-2,3-diphenylaziridine leads directly to vinylcyclopropane 111. Irradiation of this intermediate with light of wavelength 366 nm results in a unique photorearrangement that provides 112 in quantitative yield [56].

110   111   112

8 Miscellaneous

Irradiation of the 4-(3,4-pentadienyl)cyclopent-2-en-1-ones 113 with a 450 W Hanovia lamp results in intramolecular [2+2] cycloaddition to produce tricyclo[4.2.1.0^{4,9}]nonanones 114. Although the isomeric ketones 115 also result, the reaction course can be controlled by the side-chain substituents and by the temperature [57].

113   114   115

R = H,   α- or β-OH,   α- or β-OSiMe$_2$t-Bu

Wu and Houk have developed a method for constructing linearly fused triquinanes involving intramolecular [6+2] cycloaddition of fulvenes to enamines

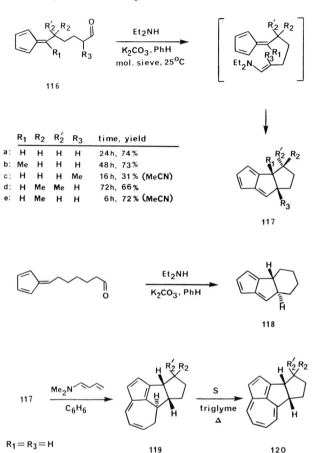

	R₁	R₂	R′₂	R₃	time, yield
a:	H	H	H	H	24h, 74%
b:	Me	H	H	H	48h, 73%
c:	H	H	H	Me	16h, 31% (MeCN)
d:	H	Me	Me	H	72h, 66%
e:	H	Me	H	H	6h, 72% (MeCN)

R₁ = R₃ = H

[58]. The conversion of 116 to 117 is exemplary and points out the exclusivety of cis ring formation. When larger rings are involved as in the case of 118, both cis and trans products result. The reaction is subject to steric hindrance; in such circumstances, a change in solvent from benzene to acetonitrile is particularly helpful. The further reaction of 117a and 117d with 1-(diethylamino)butadiene in benzene induces [6+4] cycloaddition and formation of 119. The brilliant blue tetracyclic azulenes 120 can subsequently be arrived at by refluxing 119 in triglyme in the presence of sulfur [58].

The reaction of benzvalene (121) with tetracyanoethylene and DDQ has provided direct entry to the highly functionalized diquinanes 122 and 123 [59].

121   122   123

B Ring Expansion, Contraction, and Cleavage Processes

Ring expansion of cyclobutanones to cyclopentanones is often carried out by use of diazomethane, but regioselectivity is frequently lacking. Ketone 124 is no exception [60]. On the other hand, α,α-dichlorocyclobutanones under comparable conditions undergo exclusive migration of the non-chlorinated carbon. The response of 125 is a recent example [61]. This methodology has been employed for transforming indene into 126 [62]. In an exciting development, Greene and Charbonnier have demonstrated that significant asymmetric induction can be realized in the cycloaddition reaction of dichloroketene with optically active enol ethers. The enol ether 127 derived from (-)-menthol affords 128 in 43% enantiomeric excess; use of (-)-8-phenylmenthol provides 128 in 67% optical purity. Both routes make levorotatory 129 readily available [63].

Chloro[(trimethylsilyl)methyl]ketene adds to cyclopentadiene, vinyl ethers, and silyl enol ethers to give cyclobutanones with complete regiocontrol. These products enter regiospecifically into ring expansion with diazomethane to give products that experience desilylative elimination on reaction with fluoride ion in an anhydrous dipolar solvent. Useful α-methylene ketones represented by 130 and 131 result [64].

Epoxides can be rearranged into carbonyl compounds with high efficiency by use of lithium bromide in benzene containing HMPA, and spiro epoxides derived from cyclobutanones give cyclopentanones under these conditions. New exam-

ples to surface are 132 [65a], 133 [65b], 134 [65b], and 135 [60]. Perhaps the most notable feature of the more recent work is the intriguing selectivity of bond migration observed for 133 and 134.

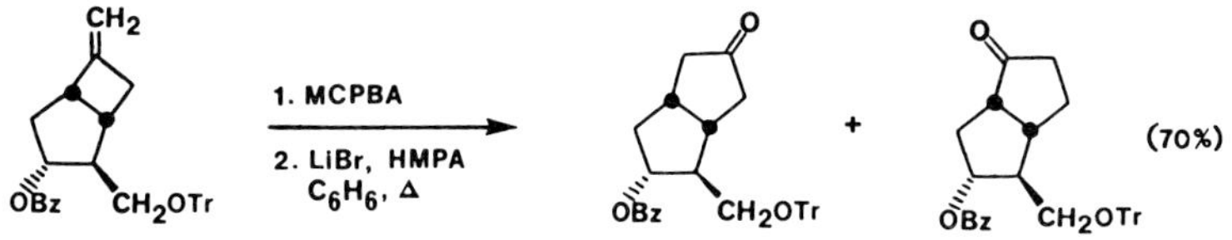

Treatment of ketones with tris(methylthio)methyllithium or tris(phenylthio)-methyllithium affords 1,2-adducts that undergo ring expansion in the presence of CuBF$_4$ or HgCl$_2$ to provide keto thioketals as exemplified by 136 and 137 (Scheme II) [60, 65]. If this procedure is followed by Raney nickel desulfurization, a serviceable route to polycyclopentanoids becomes available. This is particularly so because the ring expansion is highly regioselective, giving only the product resulting from migration of the more highly substituted carbon.

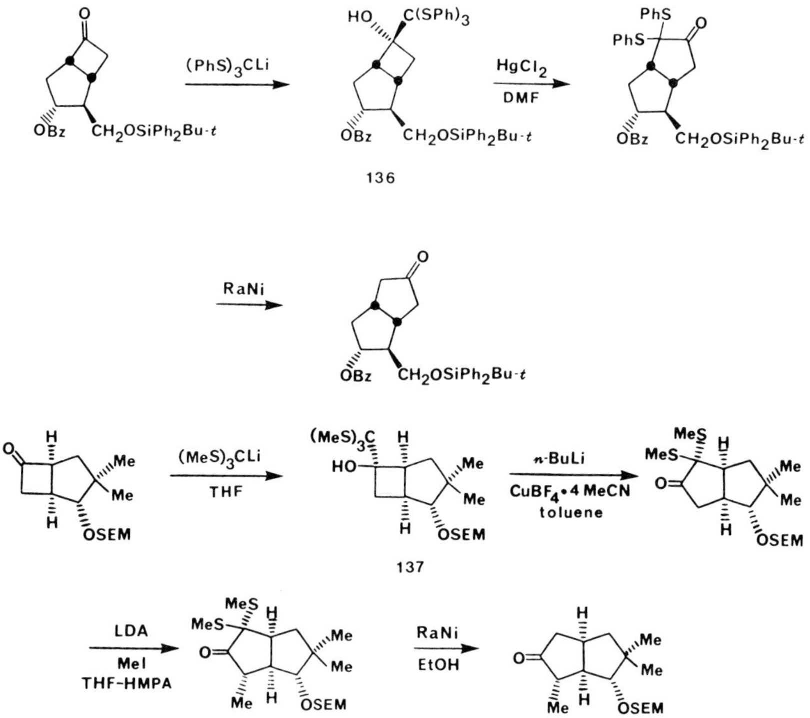

Scheme II

Murai and coworkers have made the exciting discovery that cyclobutanones are readily converted to 1,2-disilyloxycyclopentenes when exposed to a silane and carbon monoxide in the presence of dicobalt octacarbonyl [66]. The enediol disilyl ether moiety contained in 138–140 is known to be a versatile synthon.

Piers and coworkers have noted an impressive substituent effect on the site selectivity of homo-[1,5]-sigmatropic migrations that materialize during the thermal activation of vinylcyclopropanes [67]. Because the effect of a simple alkyl group on site selectivity is small, the result derived from rearrangement of 141 indicates that other molecular characteristics must be influential. The finding that 142 exhibits predominant H_x migration has been interpreted as an indication that electron density accumulates at the carbon from which the hydrogen migrates.

Barbarella and coworkers have exploited the efficient condensation reaction of 1-phenyl-2-nitropropene to keto enamines such as 143. The 1,2-oxazine N-oxides which form quantitatively in a few hours undergo isomerization to 1(2H)-hexahydropentalenone derivatives [68, 69].

Two groups have further investigated the regioselective oxidation of dicyclopentadiene to dialdehyde 144 [70, 71].

Ghosh and Saha have uncovered an abbreviated pathway for transforming indenone ketal 145 into the benzohydropentalene derivatives 146 and 147 (Scheme III) [72].

Scheme III

23

C Pauson-Khand Reaction

Cyclopentenone formation by condensation of an alkyne and an alkene with carbon monoxide in the presence of dicobalt octacarbonyl is known as the Pauson-Khand reaction [73]. The expedient formation of diquinanes is exemplified by the manner in which cyclopentadiene and 6,6-disubstituted fulvenes are transformed into 148 and 149 [74].

148 149

The intermolecular variant of this process exhibits a high degree of stereoselectivity. Also, the popularity of the reaction derives in part from the fact that structurally simple synthetic components can be defined by appropriate retrosynthetic analysis. Limitations do exist however, and these revolve in particular around the structure and/or substitution pattern in the olefinic component. The readiness with which phenylthioacetylene and *(Z)*-methyl 8-nonynoate condenses with cyclopentene to deliver 150 [75] and 151 [76], respectively, illustrates the versatility of the method.

$HC{\equiv}C-SPh$

$CO, Co_2(CO)_8$

150 (53%)

$HC{\equiv}C(CH_2)_6COOMe$

$CO, Co_2(CO)_8$

151 (33%)

Serratosa has shown that reaction of bicyclo[3.3.0]oct-2-ene with a variety of propargyl derivatives results in the exclusive formation of angularly fused triquinanes [77]. If the intermolecular organometallic cyclization is conducted at low temperature (benzene, reflux), the formation of 152 dominates. Operation at higher temperatures (140 °C) causes reduction products such as 153 to form.

152 153

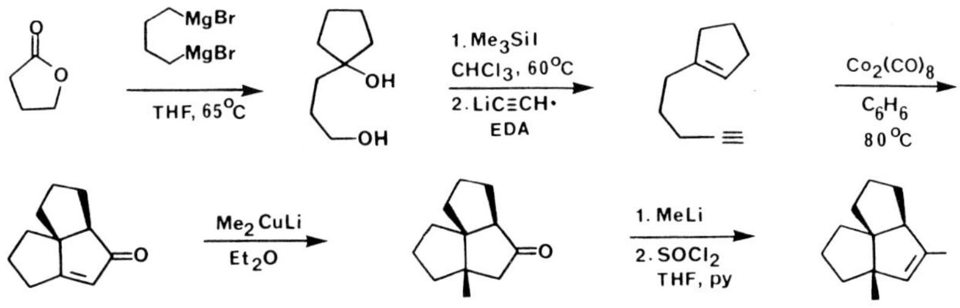

A most promising entry into functionalized polyquinanes is the intramolecular variant of the Pauson-Khand reaction. Knudsen and Schore's synthesis of tricyclic olefin 154 [78] and the preparation by the Smit group of the highly oxygenated bicyclics 155 and 156 (Scheme IV) [79] are notable because both bypass the use of carbocyclic starting materials and feature very few steps.

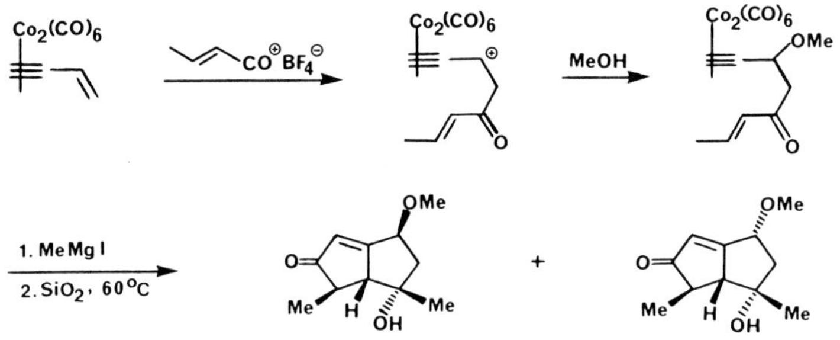

Scheme IV

In this context, it is worth noting that lactol 157 can be readily transformed into enyne 158 (Scheme V). Following conversion to the hexacarbonyldicobalt complex, heating in a sealed tube under a carbon monoxide atmosphere for 3 days

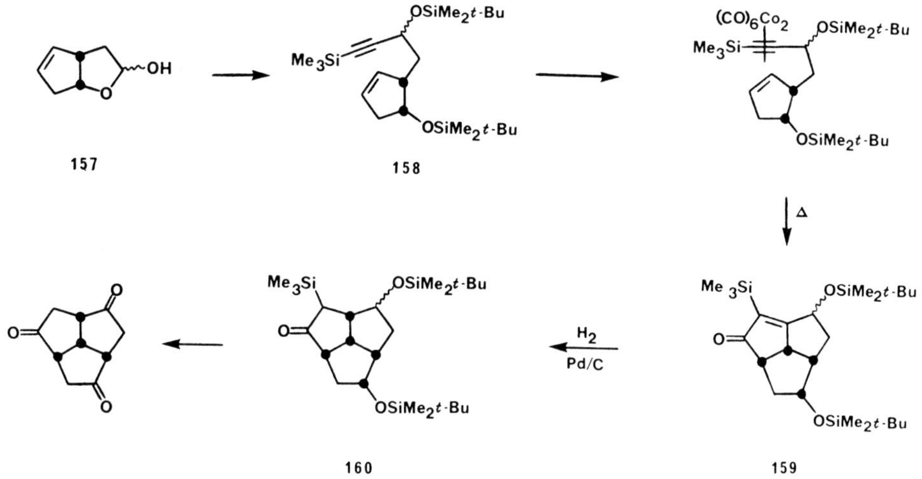

Scheme V

at 160 °C in isooctane provided 159. Subsequent hydrogenation afforded 160, the first perhydrotriquinacene to be prepared by a bisannulative Pauson-Khand process [80].

As can be seen from the behavior of 161 a and 161 b, the size of the group on the terminus of the acetylene has a significant controlling influence on the 1,3-stereoselectivity of the cyclization [81]. Magnus and Principe have also investigated the potential for 1,2-stereoselectivity in enynes 162–165 [82]. Some noteworthy facts are: (1) the silyl protecting group in propargyl system 161 is stable, but only desilylated product was isolated from the allylic system 163, and (2) the substrates 162, 164, and 165 give rise only to single, pure stereoisomers. A working mechanistic hypothesis has been advanced to explain the 1,2- and 1,3-stereoselectivity [82].

D Photochemical Approaches

1 Oxa-di-π-methane Entry

This excited-state reaction holds interest because it rapidly constructs a highly serviceable diquinane subunit. To illustrate, Diels-Alder addition of dimethyl acetylenedicarboxylate to 6-alkyl-6-carbalkoxy-2,4-cyclohexadien-1-ones such as 166 followed by catalytic hydrogenation affords 167. Triplet-sensitized photorearrangement provides the tricyclooctanones 168 [83]. Several additional examples studied by Schultz and coworkers, but based on an initial intramolecular [4+2] cycloaddition, are represented by 169–170 [84]. Clearly, the range of possibilities is large.

Electrophile-initiated ring cleavage in the parent tricyclo[3.3.0.0^{2,8}]octan-3-one is generally directed to the external cyclopropane bond, the site of maximum bond overlap [85]. The regioselectivity is, of course, enhanced in a system substituted as in 171 [86]. An additional bonus is the capacity offered for trapping the 1,3-biradical formed upon direct photochemical irradiation of these systems. The facile preparation of 172 and 173 is illustrative [86].

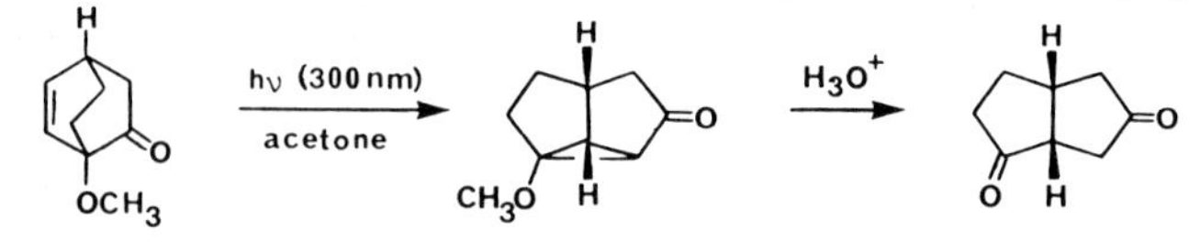

171

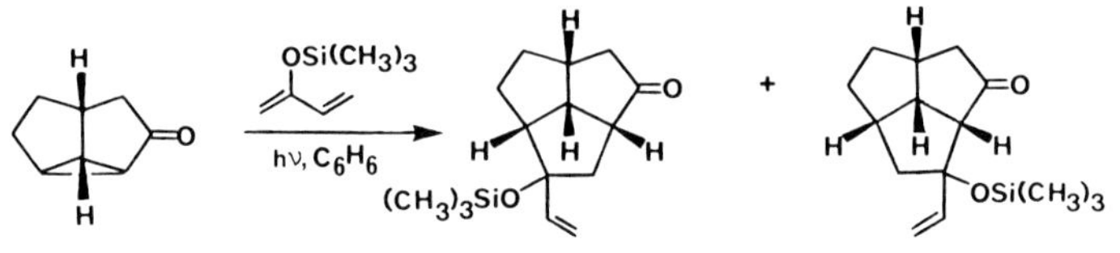

172   173

Diels-Alder addition of α-chloroacrylonitrile to bicyclic dienes 174 a and 174 b, followed by unmasking of the carbonyl group, gives rise to β,γ-unsaturated ketones 175 and 176. Irradiation of both structural types in acetone solution resulted in oxa-di-π-methane rearrangement (Scheme VI) [87]. Reductive bond cleavage in 177 and 178 proceeded with full regioselectivity to furnish the [3.3.3]propellanone 179 and the tricycloundecanone 180.

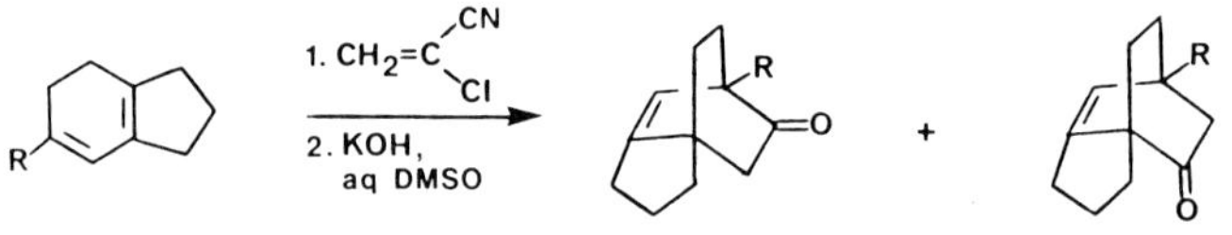

174   175   176

a, R = H; b, R = CH$_3$

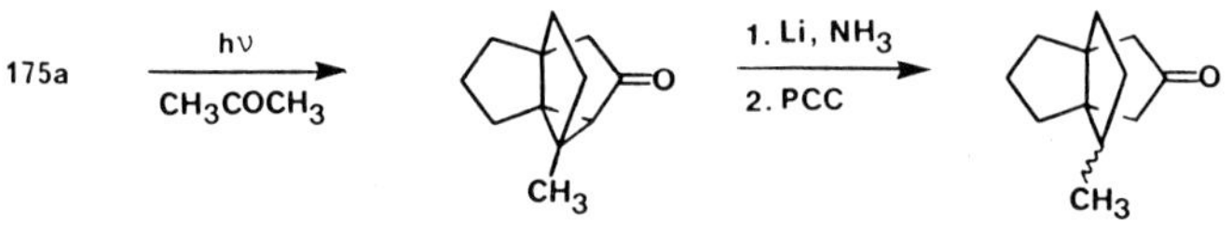

175a   177   179

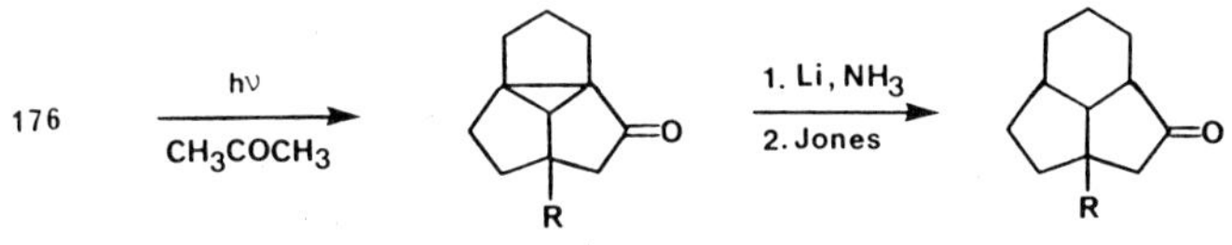

176   178   180

a, R = H; b, R = CH$_3$

Scheme VI

Callant and coworkers have used the cyclopropane ring in 181 as a temporary blocking group. Thus, formation of the dianion and addition of dimethyl carbonate gave 182. Solvolytic opening of the cyclopropane ring in 182 proceeded with concomitant decarboxylation and allowed *in situ* sodium cyanoborohydride reduction of the keto function [88]. Conversion of the resulting alcohol to 183 completed formal total syntheses of ($\pm$)-mussaenolide and ($\pm$)-8-epiloganin [89].

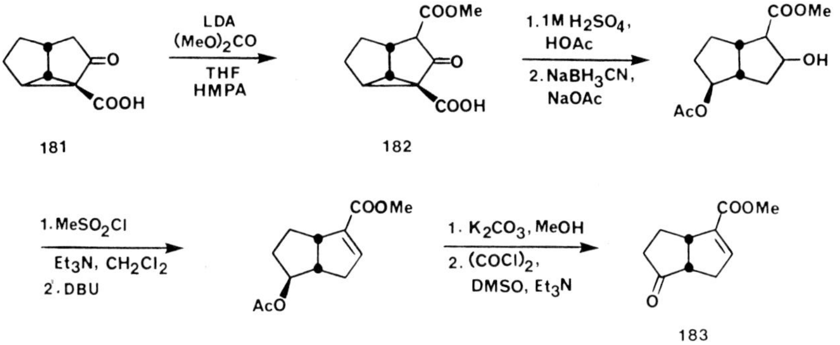

The tricyclooctanone concept has recently been extended by Demuth and Hinsken to the preparation of enantiomerically pure linear and angular triquinanes (Scheme VII) [90]. Conversion of *cis*-hydrindenone 184 to siloxydiene 185 followed by the Diels-Alder addition of maleic anhydride delivers 186. Following electrolytic decarboxylation to give 187, irradiation in acetone solution leads conveniently to 188. Application of an analogous series of steps to 189 allows access to the isomeric triquinane 190.

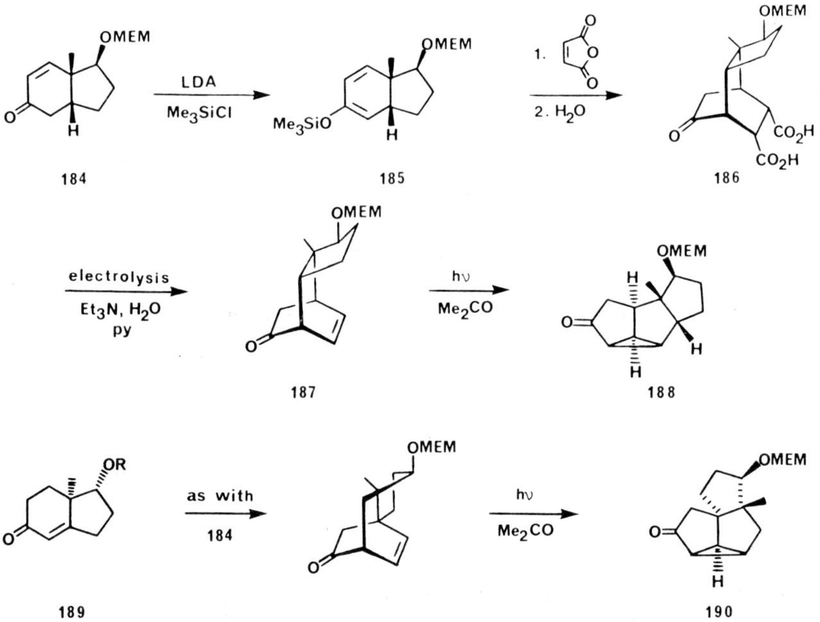

Scheme VII

Beginning with 191, it has also proven possible to prepare 192, which could serve in its own right as a potential precursor of coriolin (Scheme VIII) [91].

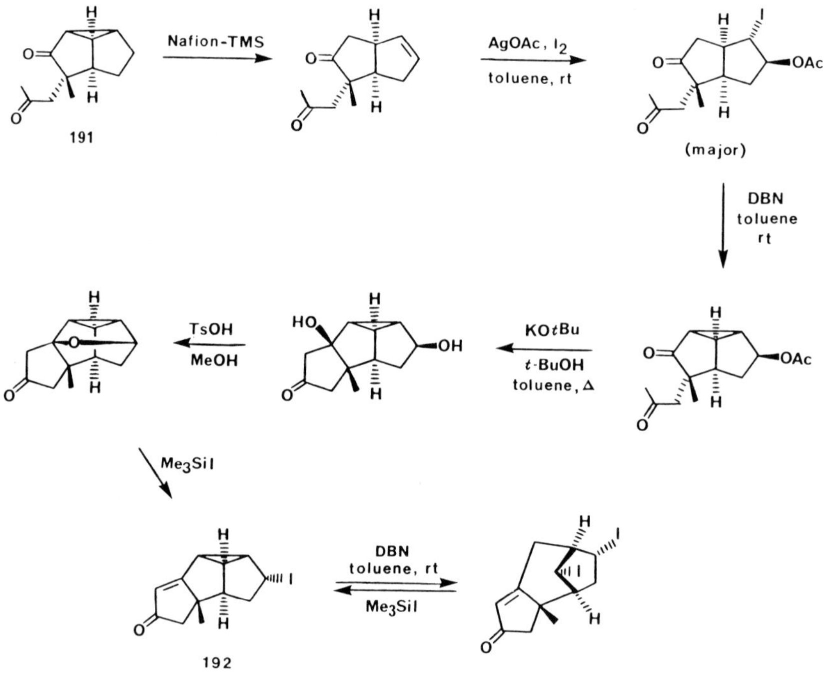

Scheme VIII

2 Meta Olefin Addition to Arenes

Reports on substituent-directed meta photocycloadditions to various olefinic acceptors continue to make frequent appearances. For example, although 1,2-dichloroethylenes are widely known to behave anomalously when irradiated in benzene, *trans*-1,2-dichloroethylene has now been found to undergo predominant conversion to 193 in the presence of benzonitrile [92]. Entirely analogous stereospecific behavior was subsequently uncovered for the three tolunitriles, α,α,α-trifluorotoluene, fluorobenzene, and chlorobenzene. Under proper circumstances, the photoadducts (e. g. 194) reacted with base to give semibullvalenes [93].

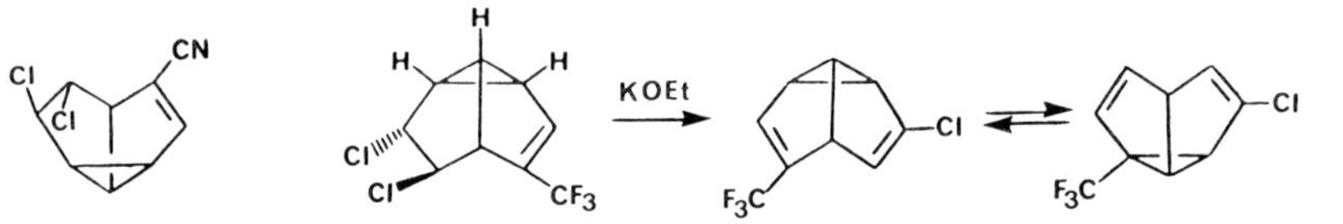

The photocycloaddition of benzonitrile and α,α,α-trifluorotoluene to cyclopentene yields principally 195–198 where the substituent resides at positions 2 or 4

R = CN
    CF₃

	195	196	197
	36%	22%	16%
	41%	27%	8%

	198		
	23%	2%	1%
	10%	14%	—

[94]. These findings have been construed by Cornelisse to agree with a reaction pathway involving polarized structures.

Added corroboration for this hypothesis has been gained from a study of the adducts formed between anisole and benzonitrile to 1,3-dioxol-2-one. The distribution of meta adducts given below is in fact consistent with the development of some charge separation along the reaction pathway [95].

A highly detailed study of the photoreactions of α,α,α-trifluorotoluene with 1,3-dioxol-2-one and several other dioxygenated alkenes has shown that charge transfer may have important consequences on the selectivity of product formation [96].

The preceding discussion underscores the degree to which the course of certain meta-photocycloadditions can be predicted. The results involving anisole and 2,3-dihydrofuran, 2,5-dihydrofuran, ethyl vinyl ether, as well as 1,3-dioxole are tabulated below [97].

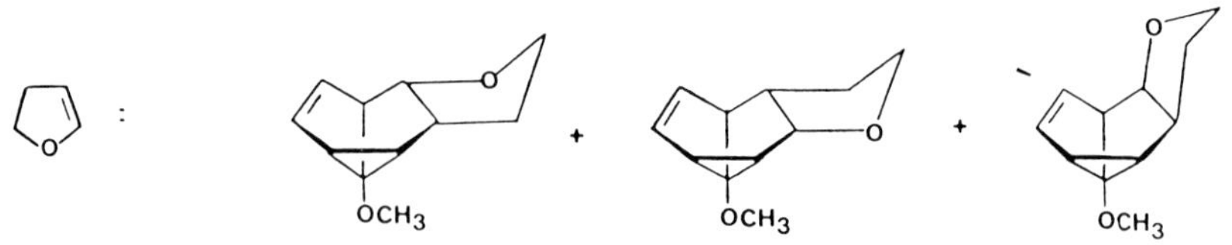

(50%)　　　　　　(33%)　　　　　　(17%)

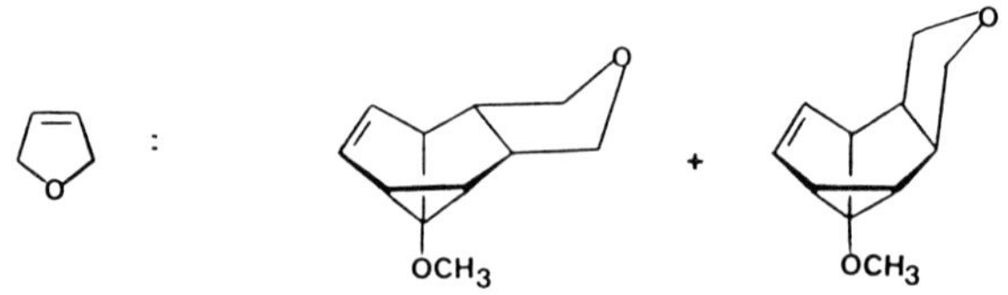

(6 %)　　　　　　(94%)

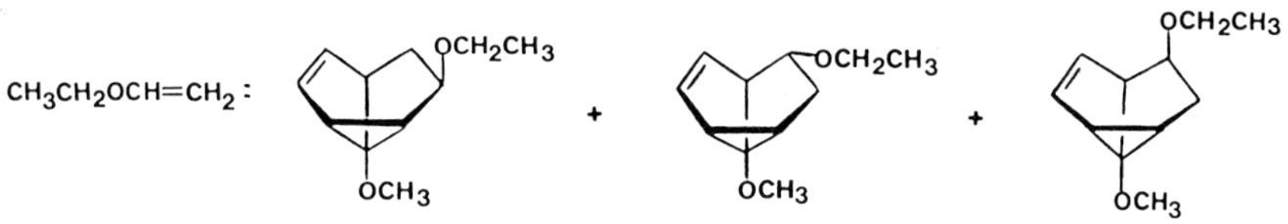

(33%)　　　　　　(44%)　　　　　　(22%)

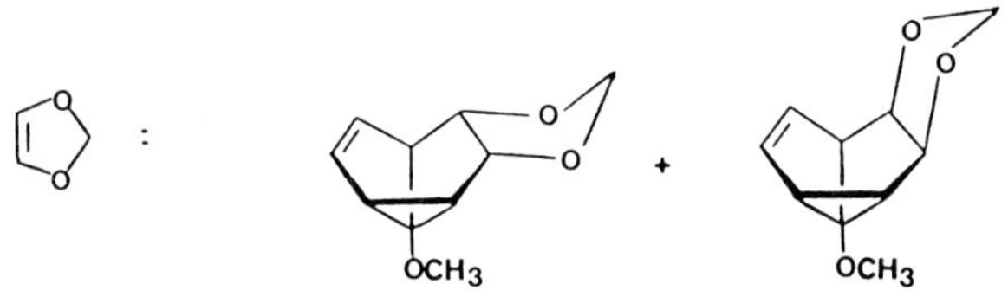

(75%)　　　　　　(25%)

The substituents in 3-cyanoanisole are positioned to work cooperatively to stabilize the dipolar intermediate. Adducts such as 199–202 with the methoxyl

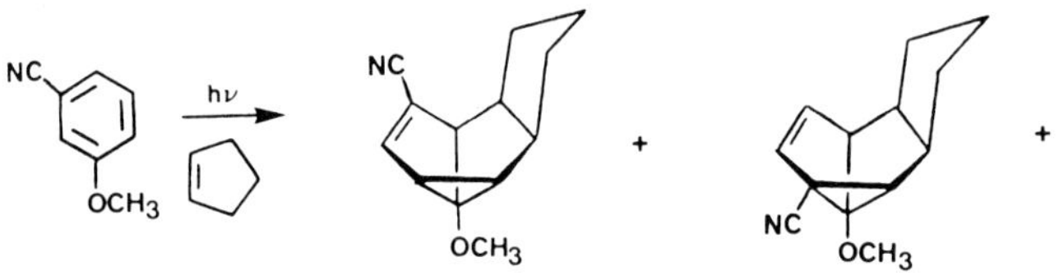

199
(40%)

200
(45%)

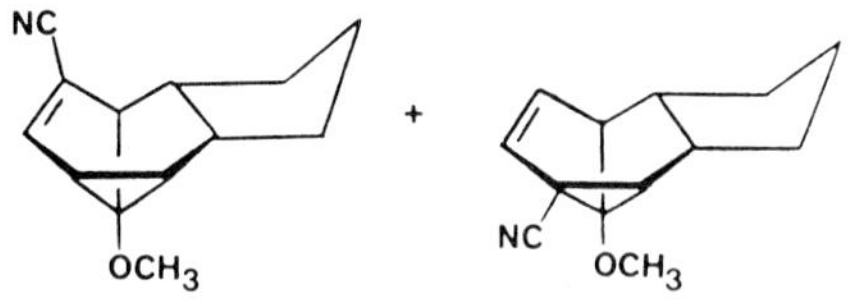

201
(5%)

202
(10%)

45%

endo (20%)
exo (10%)

endo (15%)
exo (10%)

203

group at C-1 and the cyano group at C-2 or C-4 are therefore expected and observed [98]. In 4-cyanoanisole, the intermediate cannot be stabilized simultaneously by both substituents. Since both types of products were found, it would appear that the cyano and methoxyl groups offer approximately the same amount of stabilization to the intermediate [98]. However, the formation of only 203 with *cis*-cyclooctene implicates methoxyl as the more powerful controller in this instance [99].

The meta photocycloaddition of *p*-fluorotoluene to cyclopentene yields 204 (70%), 205 (19%), and 206 (6%) [100]. Expectedly, therefore, irradiation (254 nm) of 207 in hexane produces only 208 (> 95%) [100]. The effects of methyl substitution at the 2-, 3-, 5-, and 1,1,2-positions in the pentene moiety of 5-phenyl-1-pentene have also been investigated in detail [101].

204

205

206

207

208

Secondary deuterium isotope effects have been observed in the meta cycloaddition of several aromatic compounds to cyclopentene [102]. In an attempt to prepare the putative biradicals involved in meta photocycloadditions by an

MTAD

MTAD

1. hydrolysis
2. oxidation

−N₂ → $-N_2$

$-N_2$

209   210

alternate route involving azo precursors, Sheridan prepared 209 and 210 and examined the nitrogen expulsion from these compounds [103].

E Rearrangement Routes to Polyquinanes

1 Thermochemical Pathways

Reaction of octachlorofulvene (211) with diazomethane leads to the formation of 212, which on dechlorination with stannous chloride gives 213 [104]. A spiro undecatetraene was assumed to be an intermediate in this reaction. Interestingly, Hafner and Thiele have found it possible to prepare the parent hydrocarbon (a homopentafulvalene) and to observe 214 to rearrange with stereoselective cyclopropane ring opening to give 215 at room temperature. The unsubstituted 214 can easily be prepared by oxidation of the open biscyclopentadienyl anion with cupric chloride in tetrahydrofuran at −70 °C [105]. Oxidation of the more highly substituted system 216 by this procedure furnished both 217 and 218. These two

211   212   213

214   215

isomers, which can be separated chromatographically at $-30\,°C$, rearrange faster than 214. The involvement of a diradical mechanism is suggested by the ready interconversion of 217 and 218 and by their individual conversion to 219 and 220.

Condensation of the 4-nitrobenzenediazonium ion with cyclopentadiene at $-50\,°C$ to room temperature produces the 1:2 adduct 221 and the 1:3 adducts 222 and 223 [106].

$$Ar = 4\text{-}(NO_2)C_6H_4$$

Pyrolysis of cubane at 230—$260\,°C$ affords 224 and 225 in addition to other products [107]. At $500\,°C$, 226 is quantitatively isomerized to 227 [108]. Thermolysis of enone 228 gives rise predominantly to 229 and 230 [109].

Ene and tandem Claisen-ene rearrangements continue to attract attention for the stereocontrolled synthesis of polyquinanes. Double catalysis by H^+/Hg^{++} of the intermolecular ene reaction such as 231 to 232 and 233 has been found to lower the temperature necessary to achieve cyclization [110]. Ziegler and Mikami have noted that the trimethylsilyl group can exert a profound effect on the regiochemistry of the ene reaction [111]. For example, whereas the heating of 234 provides a 1:1 mixture of 235 and 236, the silicon analogue 237 is thermally transformed only into 238. Thus, a potential method for controlling the regiochemical course of these processes may be in hand.

Rapid assembly of functionalized bicyclo[3.3.0]octanes can be achieved by means of the tandem Claisen-ene reaction. A pair of illustrative examples is given by 239 and 240 [112].

2 Cleavage of Cage Molecules

Mehta and his coworkers have pioneered the thermolysis of pentacyclic unde-
canes as a method for gaining access to *cis,syn,cis*-triquinanes. In an early
example, 241 was shown to be subject to thermal [2+2] cycloreversion to 242.
Exposure of 242 to sodium methoxide in methanol led in part to the isomerized
bisenone 243 [113]. Whereas flash vacuum pyrolysis of 244 similarly affords 245, a
precursor to 246, the ketones 247 and 248 can be arrived at by initial reductive
fission of the cyclobutane ring [114].

Methoxy groups at C-1 and C-7 noticeably accelerate the cycloreversion. A
notable demonstration of this fact is the 23-fold greater reactivity of 249 relative
to 250 [115, 116]. The presence of only one methoxyl substituent as in 251 is
adequate to permit Lewis acid catalysis of the cycloreversion at room temperature [117].

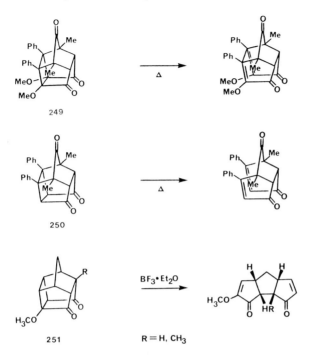

249

250

251   R = H, CH₃

The flash vacuum pyrolysis protocol has been employed to prepare the tetra-quinanedione 252 (Scheme IX) [118,119] and its mono-oxygenated analogue 253 (Scheme X) [119, 120]. A very different approach to the structurally interesting diketone has been reported by the Paquette group. Reductive sulfenylation of the domino Diels-Alder diester 254 followed by oxidative decarboxylation produced 252 (Scheme XI) [121].

252

Scheme IX

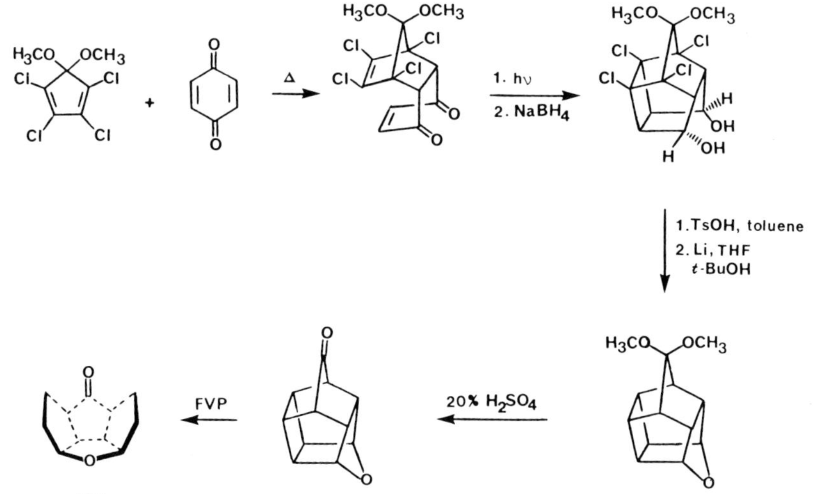

253

Scheme X

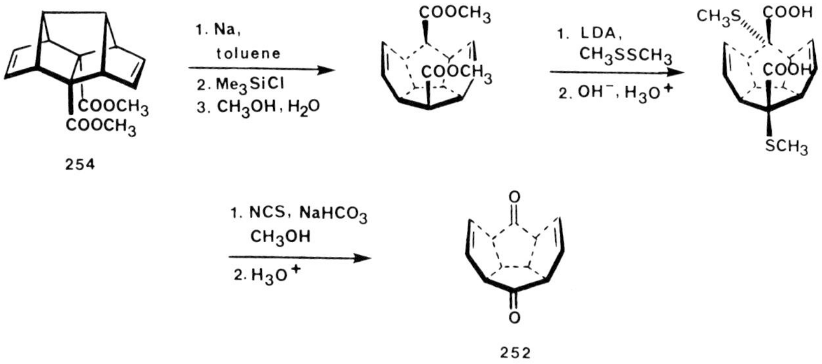

254

252

Scheme XI

241

255

257

256a, R = H
b, R = CH₃

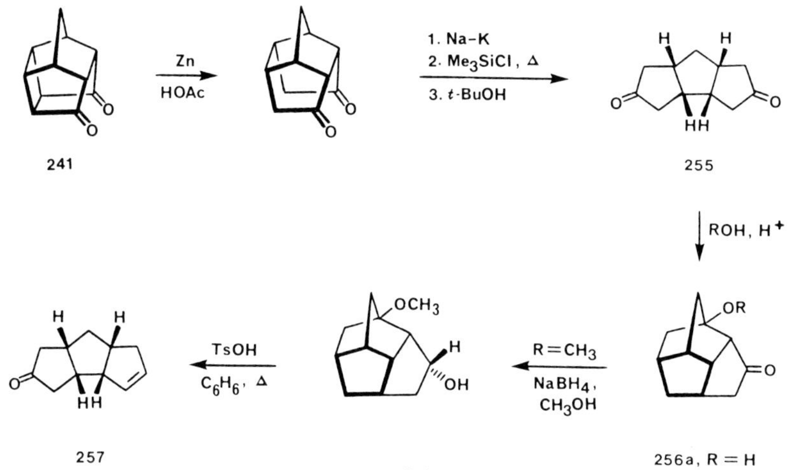

Scheme XII

Two-stage reduction of 241 leads predominantly to the formation of 255. This diketone is exceptionally prone to trans-skeletal aldol cyclization as in 256 a on contact with acid (Scheme XII). On treatment with *p*-toluenesulfonic acid in methanol, quantitative conversion to 256 b results. This tetracyclic ether permits chemo-differentiation of the two identical carbonyl groups in 255 as illustrated by its ready conversion to 257 [122].

When caged compounds 258, 259, and 260 were subjected to analogous reduction with sodium-potassium alloy in the presence of chlorotrimethylsilane and subsequently hydrolyzed in *tert*-butyl alcohol, structurally interesting poly-quinanes resulted [122, 123].

1. Na–K
2. Me₃SiCl, Δ
3. *t*-BuOH

258

as above

259

as above

260

The functionalized ketones 261 a and 261 b undergo acid-catalyzed cage fragmentation to give 262 a and 262 b, respectively [124]. Once again, the facility of this process is due to electronic participation by the bridgehead methoxyl group.

261a, X=Br
b, X=H

262a, X=Br
b, X=H

3 Carbocation-Based Approaches

Rearrangement reactions initiated by generation of a cationic center continue to provide useful means for gaining access to polyquinanes. Wilt has found, for example, that solvolysis of tosylate 263 in dioxane-water (80:20) at 110 °C results in essentially complete Wagner-Meerwein rearrangement and formation of bicycloalkene 264 [125]. Dicyclobutylidene (265) is particularly subject to formation of angularly substituted bicyclo[3.3.0]octanes under conditions of acid catalysis [126]. The symmetric diol 266 derived from cyclopentanone and acetylene is readily transformed into the spirocyclic ketone 267 under the influence of acids [127].

The pentaspirohexadecanes 268 a and 268 b may be rearranged to structurally embellished diquinanes 269 a and 269 b, respectively [128]. Acid treatment of pentaspirane 270 can lead to propellane 271 or to the hexacyclic hydrocarbon 272 depending upon reaction conditions [129]. Selected derivatization reactions of these products are illustrated.

The last couple of years have witnessed an explosion of interest in the cationic rearrangement of propellane systems. In many of the studies, polycyclopentanoid compounds play a vital role. Oftentimes as in the case of 273 and 274, the

quinanes themselves are the kinetic and/or thermodynamic termination points of the rearrangement cascade [130].

While the acetolysis of endo tosylate 275 proceeds preferentially with migration of the neighboring internal cyclobutane bond to give predominantly the isocomene-like compounds 276 (28%) and 277 (exo 3%; endo 24%), exo tosylate 278 affords the cedrene-like alcohol 279 as the major rearrangement product. Evidently, the peripheral cyclobutane bond migrates preferentially in this instance [131, 132].

The three [4.3.2]propellanes 280–282 give rise to products, the structures of which implicate operation of a concerted 1,2-shift of that cyclobutane bond having an antiplanar alignment with the leaving group [133].

42

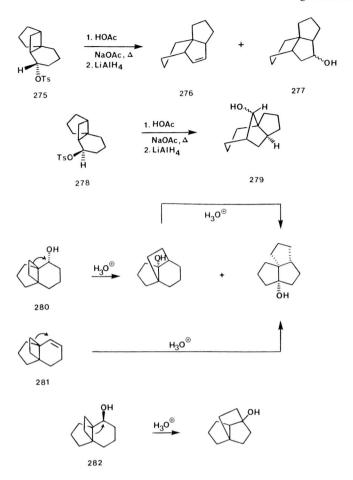

Kakiuchi and coworkers have determined that the propellanones 283–285 rearrange under the influence of p-toluenesulfonic acid in hot benzene to give the corresponding [x.3.3]propellanones. Thus, the course of skeletal isomerization, when conducted in a non-nucleophilic medium, follows an identical path regardless of the size of the third ring [134]. Reaction of 283 in aqueous tetrahydrofuran

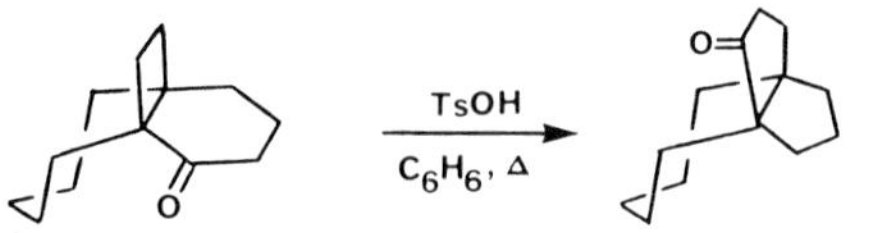

285

containing sulfuric acid affords predominantly 286 a. Similarly, treatment of 283 with *p*-toluenesulfonic acid in acetic acid yields mainly 286 b. Comparable behavior was not encountered with 284 and 285, which continued to deliver rearranged ketone ander these conditions.

The alcohol 287 has been reported to rearrange to alcohols 288 and 289 as well as hydrocarbon 276 in 50% sulfuric acid-tetrahydrofuran [132].

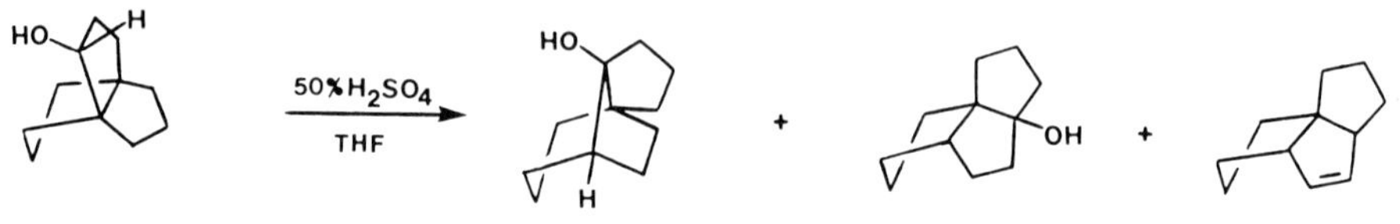

287   288   289   276

Exposure of tetracyclic ketone 290 to sulfuric acid in dichloromethane solution results in conversion to a mixture of enones 291 and 292. The same two products were also efficiently formed from the [3.3.3]propellane derivatives 294 and 295. Use has been made of this specific reaction pathway to transform 294 into decarboxyquadrone (296) [135].

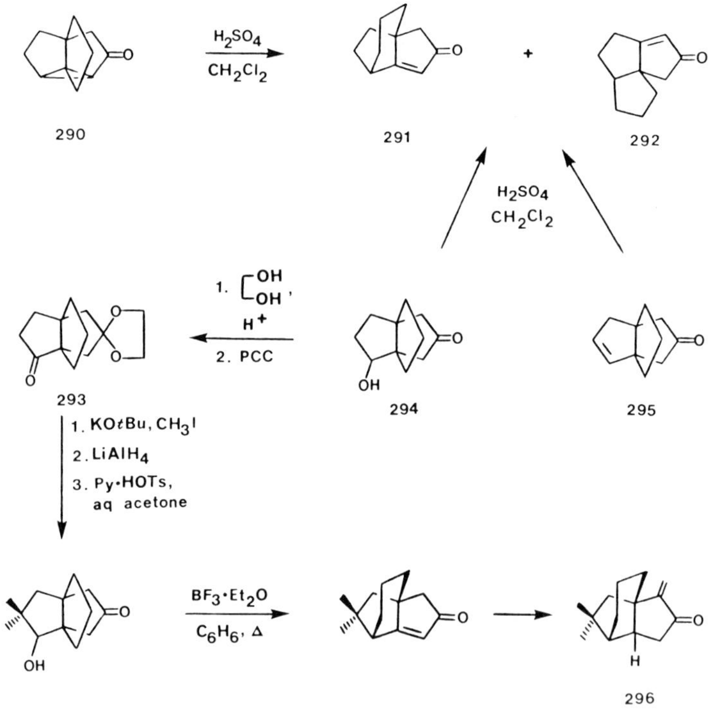

290   291   292

293   294   295

1. KO*t*Bu, CH$_3$I
2. LiAlH$_4$
3. Py·HOTs,
   aq acetone

296

The three monodeuterated adamantane isomers 297–299 were prepared as summarized in Scheme XIII. Subsequent $AlBr_3$-promoted isomerization of the individual labeled hydrocarbons to adamantane revealed 298 to exhibit an isotope effect (k_H/k_D) of 1.58. No kinetic isotope effect was operative for competition between the unlabeled 1,2-trimethylenenorbornane and 297 or 299. Consequently, the reaction must involve hydride abstraction at C-6 from the exo surface [136].

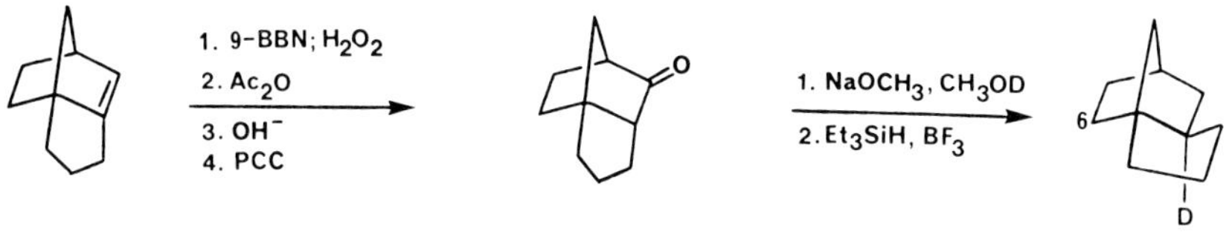

297

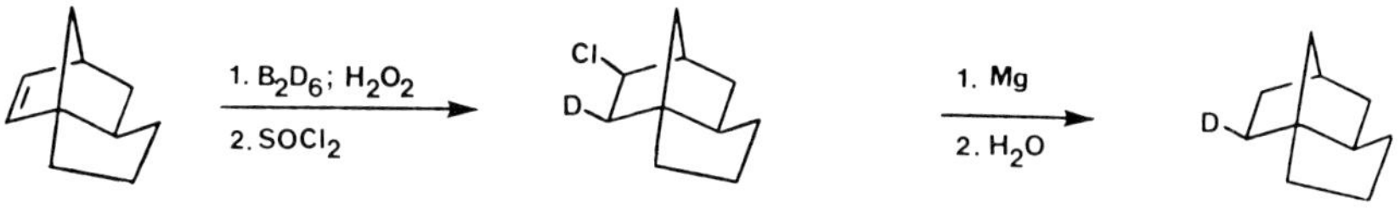

298

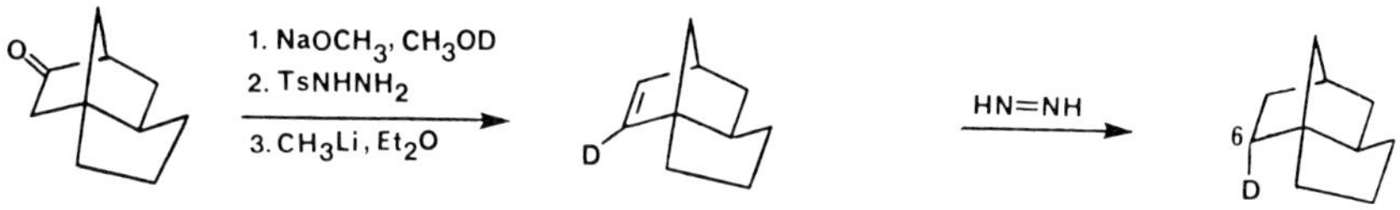

299

Scheme XIII

Natural products continue to represent prime targets for the study of cationic rearrangements, and a number of these have recently been found to eventuate in the production of polycyclopentanoid compounds. α-Cedrene (300), for example, reacts with acetic anhydride in the presence of titanium tetrachloride to afford 301 and 302 in addition to acetylcedrene [137]. Isocaryophyllene (303) is isomerized in super acid solution either to 304 or 305 depending upon the method

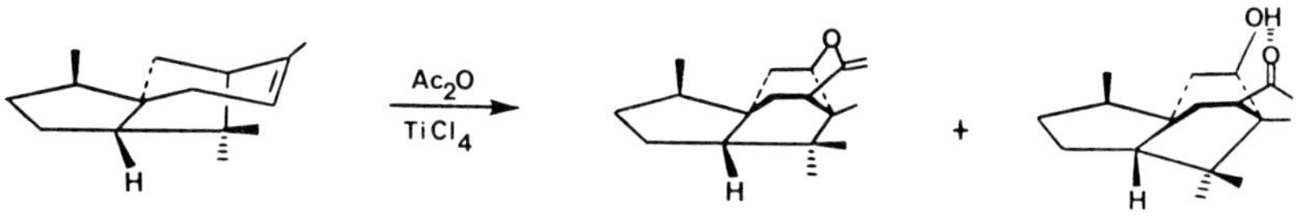

300 301 302

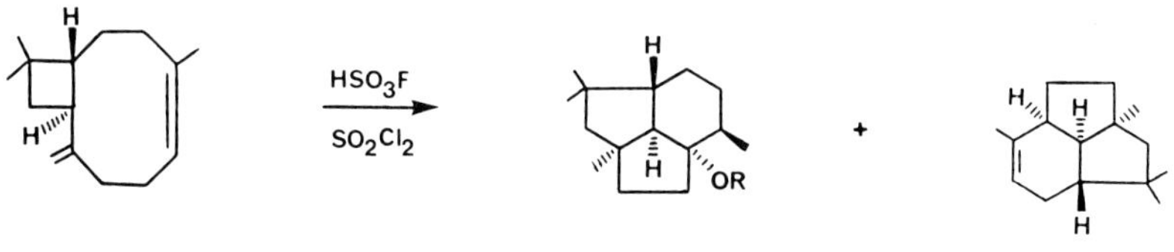

303 304 305

306   307

used for quenching [138]. A remarkable catalyst-specific rearrangement of longifolene (306) to a new tricyclic isomer termed alloisolongifolene (307) has also been uncovered [139].

The solvolytic behavior of terpenic alcohols 308a and 308b in formic acid has been studied. The two diquinanes 309 and 310 are produced in modest although different amounts [140].

308a, $R_1 = OH, R_2 = CH_3$ 309   310
   b, $R_1 = CH_3, R_2 = OH$

Bromide 311 gives 312 upon treatment with silver acetate followed by reduction as a result of exclusive migration of that C—C bond that is antiparallel to the C—Br bond [141].

311   312

4 Photochemical Processes

Direct photolysis of *cis,cis*-1,3-cyclooctadiene gives 313 among other products [142]. The major 1:1 adduct formed upon photolysis of pentamethylbenzotriazole 314 and acrylonitrile has been assigned structure 315 [143]. In solution, 315

313

314   315   316

undergoes aerial oxidation to aldehyde 316. Intramolecular [2 + 2] photocycload-dition of the allenic enone 317 affords a 3:1 mixture of the straight product 318 and the bridgehead olefin 319 [144].

Direct excitation of the polychromophoric *anti*-pentaene ester 320 with light of wavelength greater than 280 nm leads to selective formation of the nonacycle 321 (75–80%) together with lesser amounts of 322 and 323 [145]. Thermolysis of 320 at 120 °C leads uniformly to 322 [146].

When suitably constructed alkenyl tropones typified by 324 are irradiated in acidic methanol, intramolecular $6\pi + 2\pi$ cycloaddition operates successfully. The end products are bicyclo[3.3.0]octan-2-ones bridged at C 1 and C 3 with a dienyl chain [147].

In an extension of earlier work, the Frei group has noted that the presence of methyl substituents on the unsaturated ring in 325 and 326 has a beneficial effect on the vinylogous ring closure process that leads to diquinane photoproducts [148].

47

325   (58 %)   (24 %)

326   (19%)   (21%)

Irradiation of diene 327 promotes cyclization to a 3:1 stereoisomeric mixture of cyclobutenes 328 in an apparent photoequilibrium. Ruthenium tetroxide oxidation of this mixture proceeded quantitatively to furnish the α,α- and β,β-diketones 329 [149].

327   328   329

5 Base-Promoted Isomerizations

The cyclooctadienyl anion-bicyclo[3.3.0]octenyl anion interconversion continues to be the subject of investigation [150, 151]. However, it has been Stothers' application of the homoenolization principle to synthesis that has drawn recent attention. The Western Ontario group first prepared the trimethylsilyl enol ether 330 and subjected it to Simmons-Smith cyclopropanation and alkaline hydrolysis (Scheme XIV) [152]. Consistent with expectations, ketone 331 was formed and subsequently dimethylated. With 332 in hand, treatment with potassium *tert*-butoxide in *tert*-butyl alcohol at 185 °C induced homoenolization as in 333. Subsequent cleavage along path b gave the triquinane ketone 334.

330

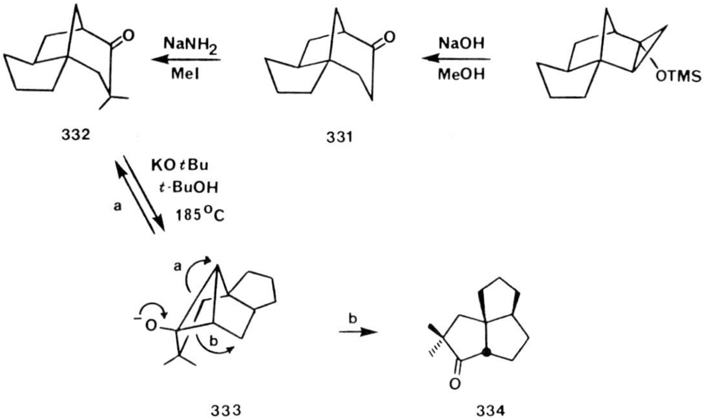

Scheme XIV

Ketone 335, prepared as shown in Scheme XV, undergoes an analogous base-catalyzed rearrangement leading to the *cis-anti-cis*-triquinane 336. The latter compound possesses a substitution plan characteristic of the hirsutane sesquiterpenes. The minor product in the β-enolate rearrangement of 335 was identified as 337 [153].

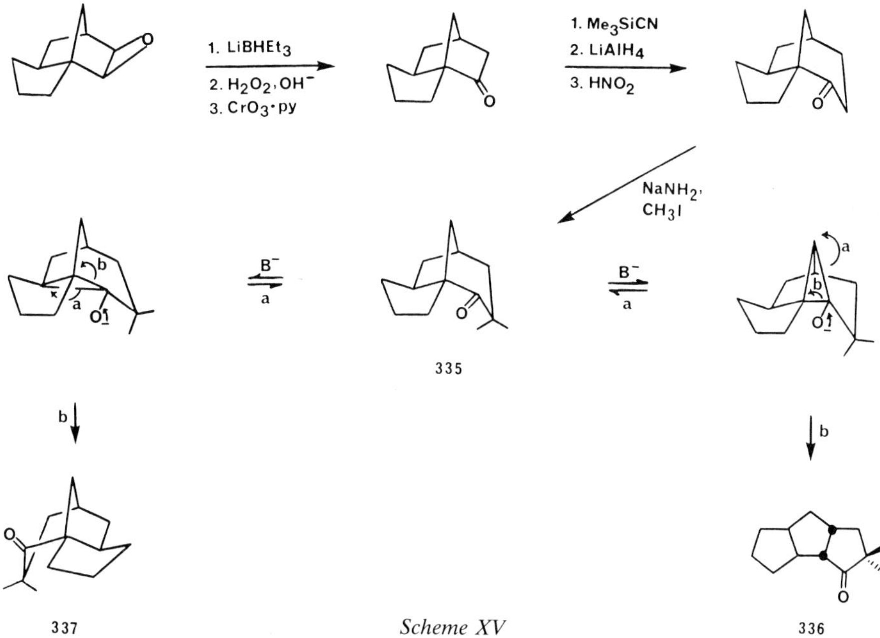

Scheme XV

More recently, the exo isomer of 7,7-dimethyltricyclo[3.3.1.0^{2,4}]nonan-6-one (338) has also been subjected to strongly basic conditions and found to rearrange readily to 339. In sharp contrast, the endo isomer is stable to these reaction conditions [154].

338   339

6 Carbenic and Transition Metal-Catalyzed Schemes

The bis-dibromocarbene adduct (340) of *cis*-hexatriene reacts regiospecifically upon treatment with methyllithium to give 1,5-dihydropentalene (341) among other products. Isotopic labelling ($\bullet$ = ^{12}C) indicated that the two bromine substituted carbons in the starting material are conjoined in the product [155].

340   341

Like its congeners, bicyclo[3.2.1]octa-2,6-dien-8-ylidene (342) is a "foiled" carbene. Decomposition of the sodium salt of its tosylhydrazone precursor gave 1,5-dihydropentalene as the major product. When the pyrolysate was immediately allowed the react with hexafluoro-2-butyne at −30 to −40 °C, the adducts 343–346 were formed. Obviously, 1,5-hydrogen shift within the dihydropentalenes is more facile than the cycloaddition reactions [156].

342

343   344   345   346

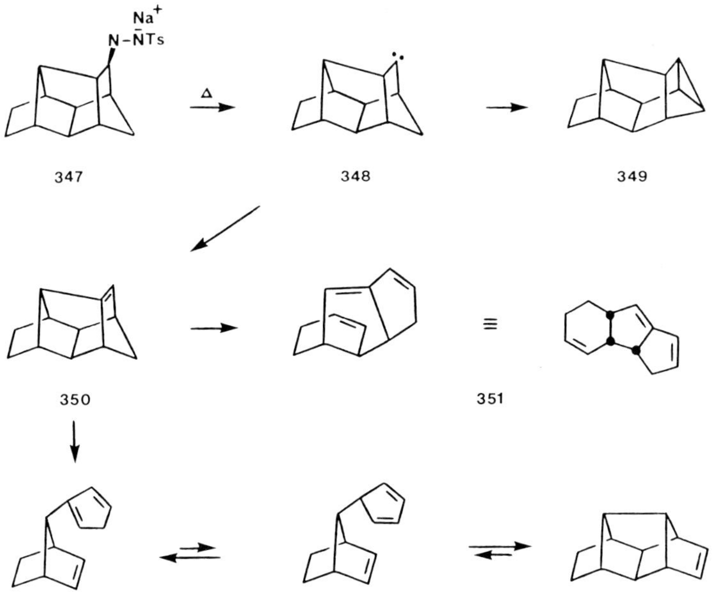

Scheme XVI

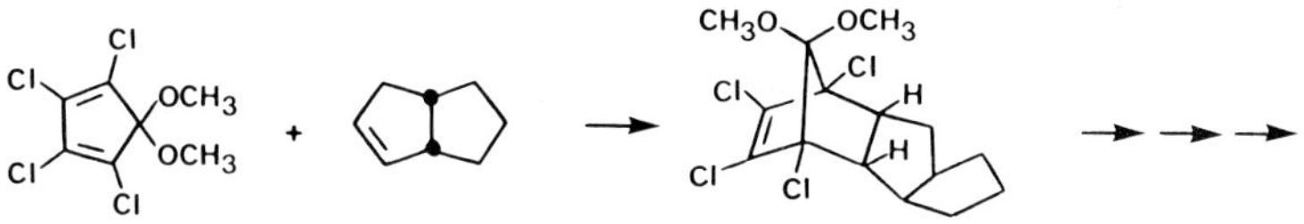

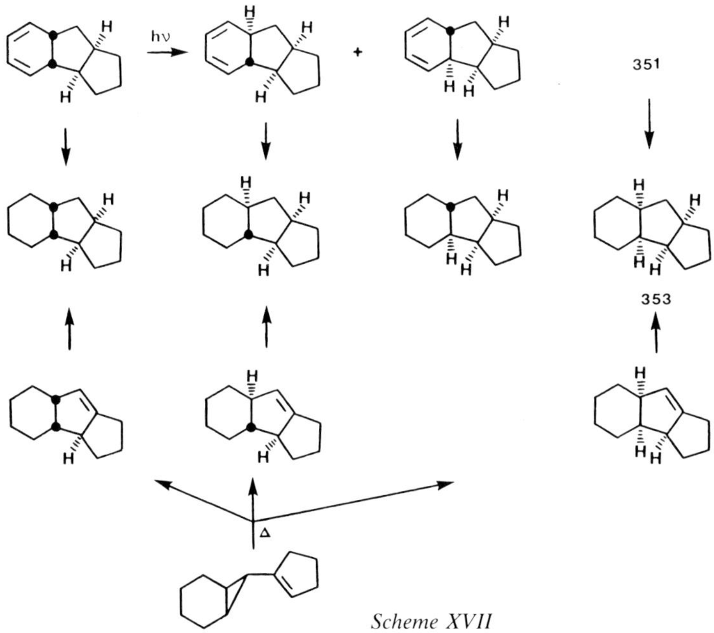

Scheme XVII

Kirmse and Ritzer have observed that carbene 348, generated by pyrolysis of 347, undergoes γ C—H insertion and conversion to 349 to the extent of only 5%. The major products 351 (33%) and 352 (61%) arise instead by competitive retro-Diels-Alder cleavage of bridgehead alkene 350 as shown in Scheme XVI [157]. Confirmation of the structural assignment to 351 was achieved by means of its exhaustive hydrogenation and independent synthesis of saturated hydrocarbon 353 as summarized in Scheme XVII.

Treatment of decachlorobicyclo[3.3.0]octa-2,6-diene (354) with tri-*n*-butyl-stannane gives 355 in low yield. Dehydrochlorination of this intermediate leads to 356, which can be prepared more expediently by dechlorination of 357 with diiron enneacarbonyl [158]. In addition, reaction of 354 with powdered selenium, sulfur, phosphorus, or titanium in the presence of anhydrous aluminum chloride at 90–100°C without added solvent results in conversion to 358, a useful prog-enitor of several chlorinated diquinane compounds.

2-Methyl-2-cyclopenten-1-ones such as 359 are produced by treating 1-vinyl-1-cyclobutanols with bis(benzonitrile)palladium dichloride and benzophenone in tetrahydrofuran solution [159].

F Trapping of 1,3-Diyls

The intramolecular 1,3-diyl trapping reaction, reknown for its ability to construct linear triquinane frameworks rapidly, is influenced by conformational, stereoelectronic, and steric factors. Asymmetric induction has now been realized in two different types of experiment. In the first, a diyl was generated that contained one stereogenic center on the chain linking the diyl and diylophile. Little and Stone prepared both 360a and 360b to determine in addition if a common 1,3-diyl intermediate would result [160]. Should this be the case, the configurational difference that is present at C—8 of these diazenes would be lost and both stereoisomers should lead to the same products and in the same ratios. In fact, this is the case. A broad range of temperatures were examined and deazatization was effected thermally and photochemically. The distribution of 361–363 was carefully assessed under the varying circumstances. Overall, the reaction is seen to generate two new carbon-carbon bonds and two new rings and to set the proper as well as absolute stereochemistry at four chiral centers.

Diazenes 364 bearing (—)-menthyl and (—)-8-phenylmenthyl ester units as chiral auxiliaries have also been prepared and similarly subjected to loss of nitrogen. Under no conditions was a synthetically useful level of diastereoselec-

tion observed [161]. In an ancillary investigation, 365 was synthesized for the purpose of evaluating the effect of a relatively short tether. The resulting diyl differs markedly in its chemistry from that of any other previously examined and produces principally 366 and 367 [162].

Substitution of a *tert*-butyl for hydrogen at the bridgehead position of 5-isopropylidenebicyclo[2.1.0]pentane changes the cycloaddition chemistry by allowing new pathways to emerge. Reaction with dimethyl acetylenedicarboxylate in degassed benzene at 25 °C does, however, result in the trapping of 368 to give 369 (25%) and 370 (24%) [163].

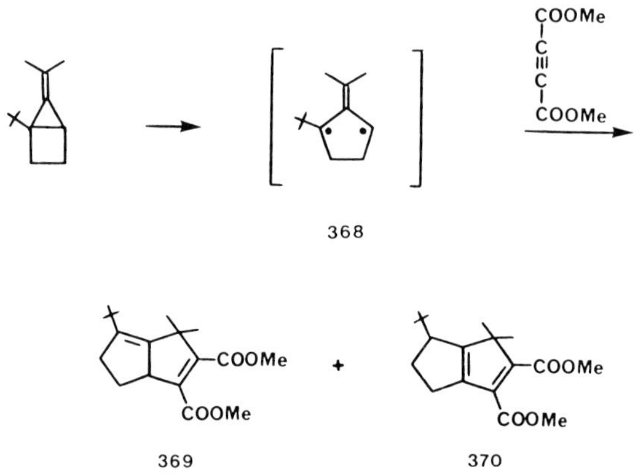

368

369　　　　　370

Until recently, the diyl trapping reaction had been restricted to diylophiles consisting of a simple C—C π bond. Little and his coworkers have now successfully extended the scope of this process to include reagents which incorporate heteroatoms. Some relevant examples are illustrated [164].

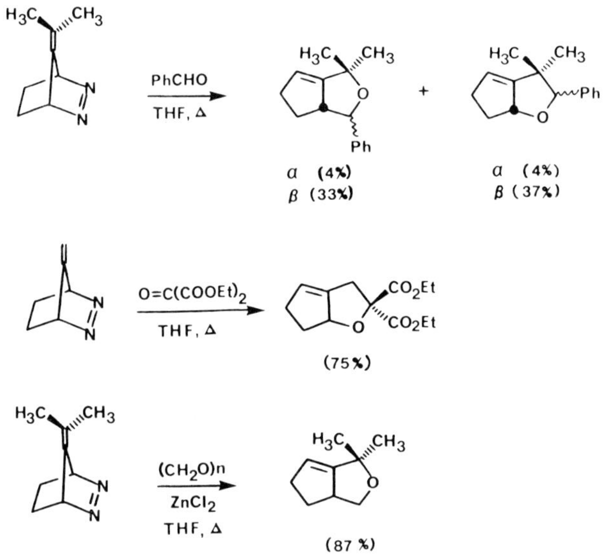

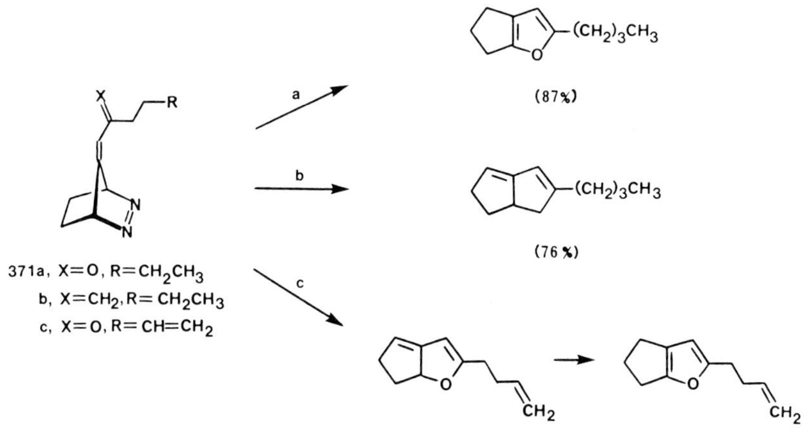

371a, $X{=}O$, $R{=}CH_2CH_3$
b, $X{=}CH_2$, $R{=}CH_2CH_3$
c, $X{=}O$, $R{=}CH{=}CH_2$

372

373
(70%)

For comparative purposes, the three diazenes 371 a–c were thermally activated. In all cases, cyclization occurred smoothly. Initially formed 372 rearranged to 373 during chromatography over silica gel [165].

G Transannular Cyclizations

The cyclooctadienyne 374 has been found to isomerize rapidly to 1,2-dihydropentalene under flash thermolysis conditions (450–640 °C, 0.3 torr) [166]. Reports have continued to appear demonstrating the extent to which 1,5-cyclooctadiene can form functionalized diquinanes when treated with various reagents. Thus, hydrophosphorylation [167], iodination in carbon tetrachloride [168], iodination in acetonitrile [168], and the sulfur trioxide-promoted addition of N,N-dialkyl-

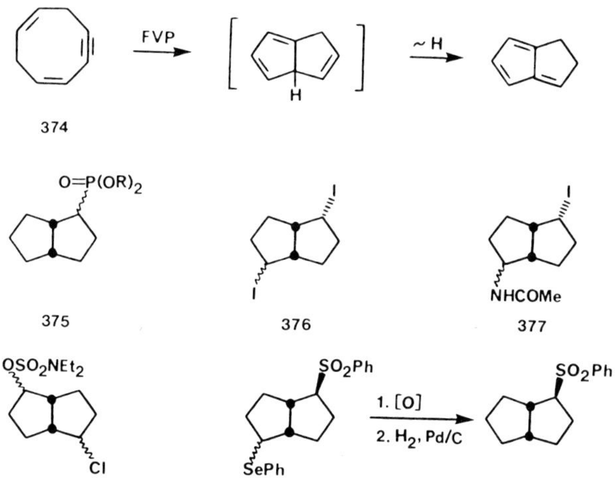

374

375 376 377

378 379 380

chloramines [169] give rise to 375–378, respectively. Selenosulfonation leads to 379, a serviceable precursor to sulfone 380 [170].

The diene 381 reacts rapidly with one equivalent of bromine to give dibromide 382 [171], the three-dimensional structure of which has been elucidated by X-ray analysis [172].

381   382

Reduction of 5-iodo-1-cyclooctene (383) with lithium triethylborohydride results in the formation of bicyclo[3.3.0]octane, presumably by a single electron transfer process [173]. Diepoxide 384 in the presence of lithio ethylenediamine is transformed to a minor extent into 385 [174]. Hydrophosphorylation of cis,trans,-trans-1,5,9-cyclododecatriene leads to three products, two of which have diquinane part strucures [167].

383

384   385   (91.5%)

(8.5%)

Intramolecular reductive coupling of 1,5-cyclooctanediones with low-valent titanium species has proven useful for the elaboration of polyfused cyclopentanoids [175, 176].

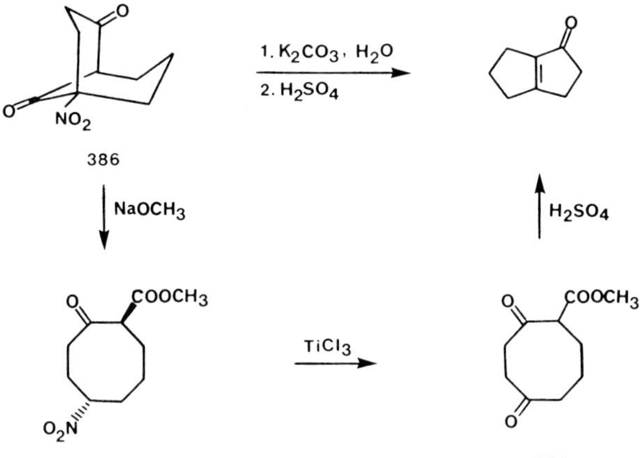

386 387 388

The unusual conversion of nitro diketone 386 to bicyclo[3.3.0]oct-1(5)-en-2-one has been achieved under two very different sets of conditions [177]. The first simply involves sequential treatment with aqueous potassium carbonate and then sulfuric acid. The second pathway is mediated by the functionalized cyclooctanes 387 and 388.

A carbonium ion-mediated transannular cyclization of the suitably functionalized bicyclo[6.3.0]undecane 389 has been employed as the pivotal step in the construction of angularly fused triquinane 390 (Scheme XVIII) [178].

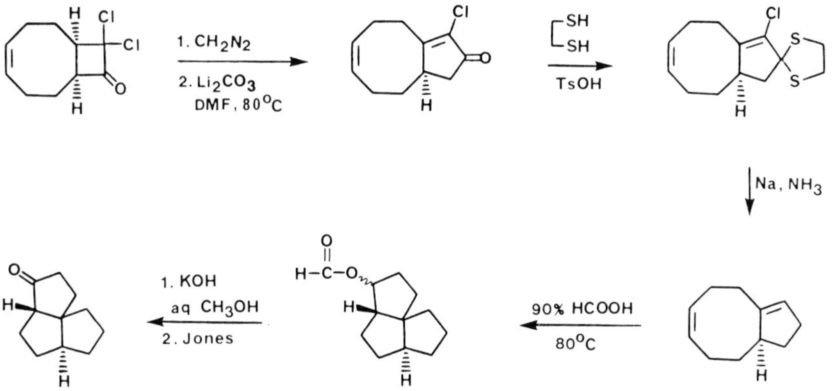

390 389

Scheme XVIII

III Functional Group Manipulation Within Polyquinanes

A Reactions Involving Ketonic Substrates

An attempt to prepare 392 by intramolecular C—H insertion within the carbene generated from 391 led instead to 393 [179]. The conversion to anhydride 394 was successful, and this propellane anhydride was also prepared from two other precursors.

Reduction of 395 with sodium borohydride afforded the three possible diol configurational isomers [180]. Stepwise reduction of the higher carbocyclic propellanedione 396 was realized via monoketal 397 [181]. Vinyl triflate 398 has

been prepared by reduction of the α, β-unsaturated ketone with L-Selectride and trapping of the resulting enolate with N-phenyltriflimide [182]. Treatment of bicyclo[3.3.0]oct-7-en-2-one with dianion 399 followed by acid-catalyzed hydrolysis and cyclodehydration in a "one-pot" operation provides the spiro(E)-α-ethylidene-γ-butyrolactone 400 [183].

Reaction of ketones 401a and 401b with cyclopropyldiphenylsulfonium tetrafluoroborate and potassium hydroxide produces oxaspiropentanes, which afford the vinylcyclopropanols on exposure to lithium diethylamide. As shown in Scheme XIX, treatment of these intermediates with one equivalent of bromine-dioxane complex or *tert*-butyl hydroperoxide in the presence of a vanadium catalyst produces a mixture of 402 and 403, or 404. Baeyer-Villiger oxidation of 404 with basic hydrogen peroxide led to the hydroxy lactones 405a and 405b [184].

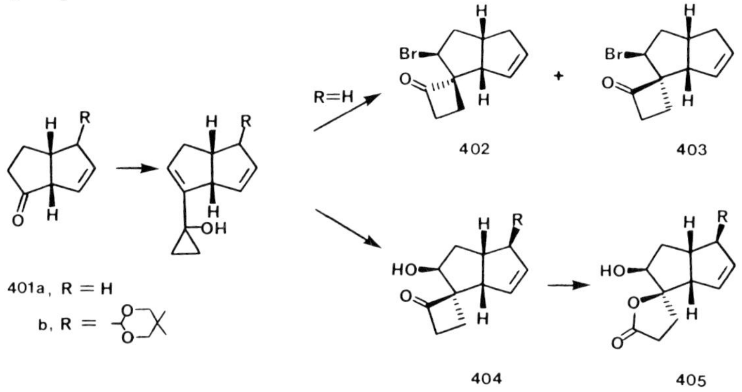

Scheme XIX

A similar protocol has been developed to achieve C-alkylation α to the carbonyl. The formation of 406 constitutes one example [185].

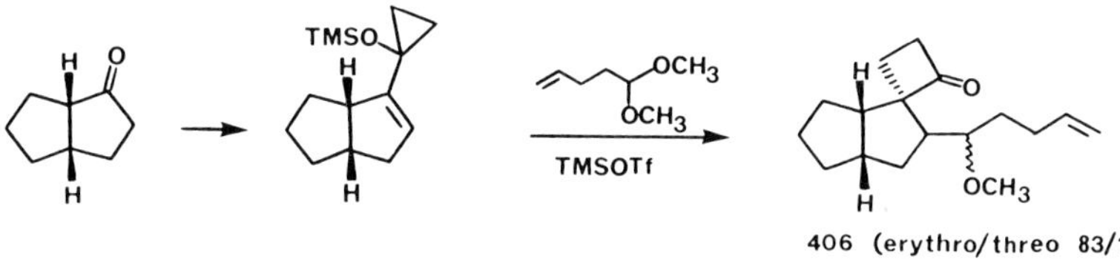

406 (erythro/threo 83/17)

Diethyl dicarbonate has proven to be a convenient reagent for the preparation of 407 and related β-ketoesters [186]. O-Silylated ketene acetals have been added to activated enone 408 under 15 kbar of pressure in acetonitrile at 20 °C [187]. Treatment of lithium enolates with trialkylstannyl trifluoroacetates permits their use as nucleophiles in allylic alkylation with 409 when promoted by palladium(O) [188].

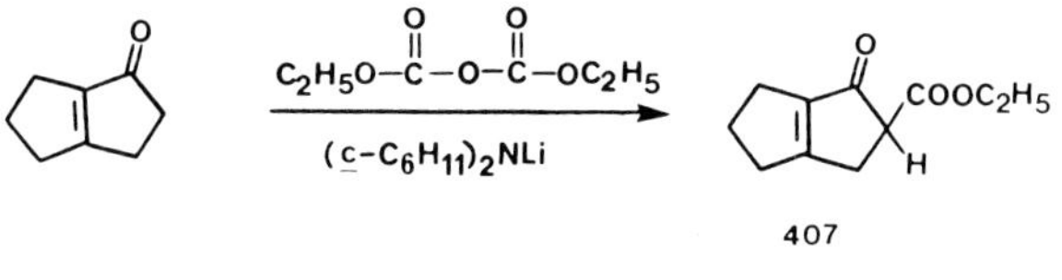

407

Oxidation of 2-(phenylthio)bicyclo[3.3.0]octane-3,7-dione (410) with *m*-chloroperbenzoic acid gave rise to a 1:1 mixture of sulfoxides 411 and 412 whose configurations were assigned on the basis of their relative rates of elimination to 413 at 100 °C [189]. Ketalization prior to thermal extrusion of phenylsulfenic acid provides 414 [190]. This versatile diquinane building block has also been prepared from 415 and shown to undergo ready 1,4-addition with cuprate reagents [191].

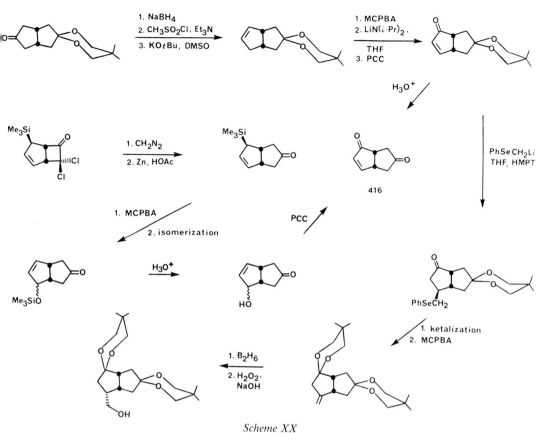

Scheme XX

cis-Bicyclo[3.3.0]oct-3-ene-2,7-dione (416) has been elaborated by two different methods (Scheme XX) and shown to function well as a Michael acceptor [192].

Jähne and Gleiter have transformed 1,5-dimethylbicyclo[3.3.0]octane-3,7-dione into the tri- and tetracyclic counterparts 417 and 418 by sequential bromination and transannular dehydrobromination [193].

Upon irradiation in acetone, 2-(2-propenyloxy)cyclohex-2-enone (419) gives 420. Hydride reduction of this head-to-tail photoadduct to the α-alcohol, mesylation, and cationic rearrangement has been observed to give 421, acid hydrolysis of which delivered the functionalized diquinane 422 [194]. In comparable fashion, irradiation of 423 has afforded 424, which was transformed into 425 [194].

Jones oxidation of 426 delivers keto lactone 427 which when allowed to stand in the presence of potassium acetate undergoes conversion to enone 428. Sequential reaction of this intermediate with carbonyldiimidazole and the magnesium monomalonic ethyl ester complex gave rise efficiently to 429, thereby achieving oxa → methylene transposition [195].

The first step of an unusual approach to the total synthesis of 9,11-dehydroestrone takes the form of a simple alkylation of tricyclo[3.3.0.0^{2,8}]octan-3-one

430

432 **431**

to give 430. The action of acetyl methanesulfonate on 430 results in cyclopropane ring cleavage and formation of 431. During this reaction, spontaneous partial hydrolysis occurs and direct closure to 432 is realized in >70% yield [196].

The Dewar-type isomers of azulene and methoxyazulene (437 and 439) have been synthesized from the common intermediate 435 [197]. Introduction of an α, β-double bond into 433 was effected with sodium hydride and methyl p-toluenesulfinate followed by thermal elimination. Irradiation of 434 in a E/Z-mixture of 1,2-dichloroethylene afforded tricyclic chloride, which was transformed into 435 and 436 (ratio 9:1) by means of acetalization, reductive elimination, and deacetalization. Bromination of 435 with N-bromosuccinimide afforded a mixture of allylic bromides, direct treatment of which with potassium *tert*-butoxide in HMPA and then CH$_3$OSO$_2$F gave 437 as an extremely air-sensitive yellow oil. If 438 is instead reduced with sodium borohydride, an epimeric mixture of alcohols results. Exposure of this mixture to tri-*n*-butylphosphine produced 439.

433 **434** **435** **436**

439 **438** **437**

Although attempts to effect the Wharton rearrangement in 440 were to no avail, it proved possible to effect alkylative enone transposition by Grignard addition and subsequent PCC oxidation [198].

Dichloroketene addition to 257 proceeds stereoselectively from the convex face but not regioselectively to give a nearly 1:1 mixture of 441a and 441b. Diazomethane ring expansion and reductive dechlorination furnished tetraquinane diketone 442. Epoxidation of 257 yielded the exo-epoxide 443, a molecule very subject to base promoted intramolecular opening and formation of 444 [123].

B Carbocationic Processes

The bicyclic alcohol 445 undergoes the Ritter amidation reaction to give amide 446, which can also be produced by Beckmann rearrangement of 447 [199]. Koch-Haaf carboxylation of 445 correspondingly leads to carboxylic acid 448. In contrast, the tertiary alcohols 449a and 449b furnish the unisomerized carboxylic acids 450 [200].

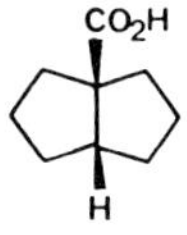

448 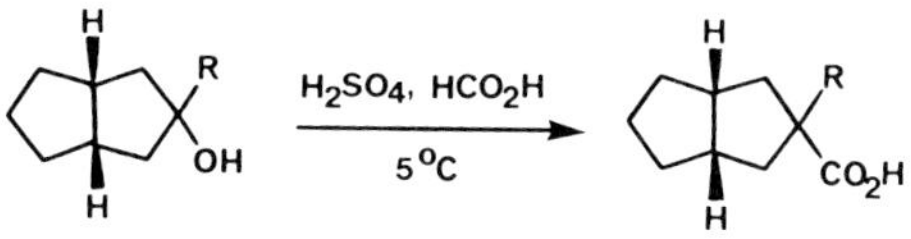   449   450

a, R = CH₃; b, R = C₆H₅

Brown has published the complete details surrounding the solvolytic behavior of secondary and tertiary systems of type 451 and 452. In particular, the progressive increase in exo/endo rate ratios with increasing U-shaped character is given extensive discussion. Steric retardation of ionization of the endo isomer is believed to be a major factor in governing the exo/endo ratios, particularly in the tertiary systems [201, 202].

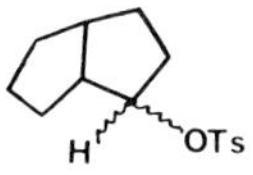

451
exo/endo = 1.3

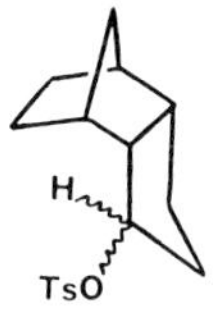

exo/endo = 5.7

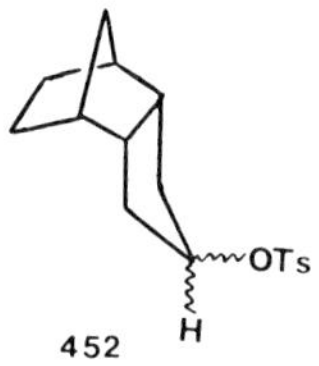

452
exo/endo = 0.6

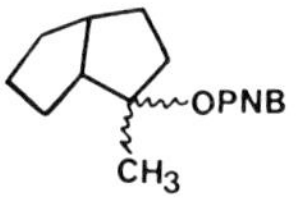

exo/endo = 17

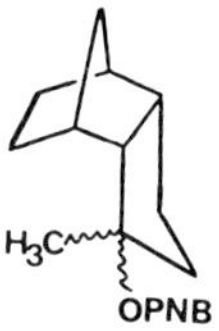

exo/endo = 4300

Hydride reduction of amide 453 affords the aminomethyldinoradamantane 454. Heating this substance with sodium nitrite in acetic acid provides alcohol 455. This unexpected Demjanow rearrangement has been rationalized in terms of consecutive skeletal reorganizations via the twist-brendanyl cation 456 [203].

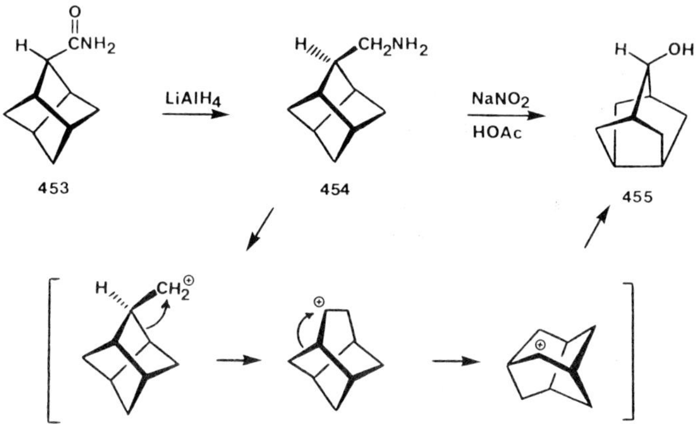

453   454   455

456

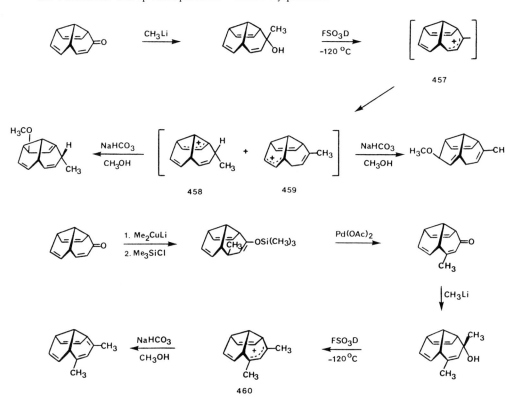

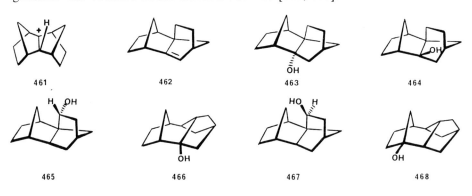

Whereas cation 457 is a short-lived intermediate and experiences hydride shift to the thermodynamically more stable entities 458 and 459, its homologue 460 is contrastingly stable [204]. However, no homoaromatic character could be detected in this system by NMR spectroscopy.

The nature of cation 461 has been probed both solvolytically and by direct NMR examination. This most sterically congested species is produced in aqueous acetone 3,500,000 times more rapidly than the *tert*-butyl cation and shows no tendency for skeletal rearrangement [205]. In contrast, fusion of two norbornyl systems in the alternative manner depicted in 462 introduces a propensity for a cascade of cationic rearrangements. Depending upon acid strength, one can generate one or more of the alcohols 463–468 [205, 206].

461   462   463   464

465   466   467   468

Although the two triepoxide stereoisomers 469 and 470 rearrange smoothly under the influence of Lewis acids or heat to the topologically nonplanar trioxahexaquinane 471, the same involvement of the ensemble of three-membered rings in 472 and 473 has not been realized. Whereas 472 gives rise to a myriad of unidentified products, 473 isomerizes only to aldehyde 474 in the presence of anhydrous zinc chloride [207].

469   471   470

472   473   474

C Reactions Involving Olefinic Centers

The efficiency with which enone 475 undergoes organocuprate addition has been documented [208], as has the enecarboxylation of *cis*-bicyclo[3.3.0]oct-2-ene with diethyl oxomalonate to give 476 [209]. Reaction of chiral glyoxylate 477 with diene 478 at −78 °C in the presence of stannic chloride affords 479 and offers a highly attractive means to achieve breaking of the molecular symmetry inherent in 478 [210].

475   476

477   478   479

Serratosa, et al., have examined the capability of the acetal groups in 480 and 481 to direct the course of oxymercuration-reduction and hydroboration-oxidation. As anticipated by them, hydroboration of 480 exhibited no regioselectivity;

	480 →		:	
BH$_3$•THF		50	:	50
TxBH$_2$		50	:	50
9−BBN		50	:	50
(Sia)$_2$BH		50	:	50
Hg(OAc)$_2$		10	:	90

	481 →		:	
BH$_3$•THF		62	:	38
BH$_3$•SMe$_2$		61	:	39
TxBH$_2$		62	:	38
9−BBN		64	:	36
(Sia)$_2$BH		85	:	15
Hg(OAc)$_2$		30	:	70

for 481, a marked preference for formation of the 1,3-isomer was seen. The consequences of oxymercuration have been attributed to coordination of Hg^{+2} to the nonbonded electron pairs of the endo-oxygen atom of each acetal group [211].

The Deslongchamps triquinacene intermediate 482 can be conveniently transformed into either of the two isomeric dienones 483 or 484 depending upon conditions [212].

Optically active [m.n.l]propellanones typified by 485 have been prepared by diastereoselective cyclopropanation of homochiral ene ketals derived from 1,4-di-O-benzyl-L-threitol [213].

The readily available diketone 242 has served as the starting material for a synthesis of precapnelladiene [214]. Relocation of one of its double bonds and partial hydrogenation gave 486. Selective Wittig olefination and stereoselective hydrogenation produced 487 which was deoxygenated via a thioacetalization-desulfurization sequence. The resulting olefin 488, when oxidized with catalytic ruthenium dioxide under Sharpless conditions, afforded the pivotal 5/8 bicyclic diketone 489.

An attempt to realize anionic bis[2,3]sigmatropic rearrangement in 490 has been shown to be circumvented by an $S_N i'$ displacement producing tricyclic ether 491 [215].

(R = H, CH₃)

Reaction of dilithiopentalene (492) with methyllithium and methylene chloride leads *inter alia* to the interesting diquinane hydrocarbons 493 (38%), 494 (6%), and 495 (2%). Diimide reduction of the latter afforded 496, an alternative synthesis of which was developed from 478 [216].

D Miscellaneous

Alcohols 497 and 498 are chemoselectively oxidized to their ketones with hydrogen peroxide in the presence of ammonium molybdate and potassium carbonate [217]. Carboxylic acid 499 has been resolved and its absolute configuration assigned [218]. The bicyclo[3.3.0]octadienediyl dianion (500) undergoes ring opening to give the cyclooctatetraene dianion; this behavior is the reverse of that observed for the corresponding dication [219].

Thermal decomposition of optically active 501 yields optically active dihydrosemibullvalene [200, 221]. Comparable denitrogenation of 502 and 504 leads to semibullvalenes 503 and 505, respectively, as major products [220]. These results indicate that the loss of nitrogen proceeds via a concerted pathway (see 501).

Cyclopropyl keto ester 506 reacts smoothly with lithium dimethylcuprate to furnish 507 [222]. Treatment of the γ-hydroxyalkyl stannanes 508 and 509 with lead tetraacetate in refluxing benzene results in ring cleavage and stereospecific formation of the new double bond [223].

508

509

IV Physical Data for, and Theoretical Analysis of, Poly-quinanes

A Crystal Structure Data

The two 3,7-disubstituted diquinanes 510 and 511 have been investigated by single crystal X-ray diffraction techniques and shown to possess the illustrated W conformation [224]. Nonbonded H----H distances of 1.754(4) and 1.713(3) Å have been measured by low-temperature neutron diffraction experiments for 512 and 513 [225]. At 2.921(4) Å, the distance between the pairs of double bonds in 514 is also quite short [226].

510

511

512

513

514

Single crystal X-ray analyses have been carried out on the three isodicyclopen-tafulvenes 515–517. Two important features emerged: (a) the angles about the apical methano carbon of the norbornane ring (96.2—96.7°) exhibit the usual deviation from a tetrahedral value; (b) the dihedral angles between the bridge-head C—H bonds and exocyclic π bonds in all three molecules are not significantly different [227]. Despite the fact that the three fulvenes are almost indistinguishable in their overall structural topology, they exhibit quite different π-facial stereoselectivity toward dienophiles.

515

516

517

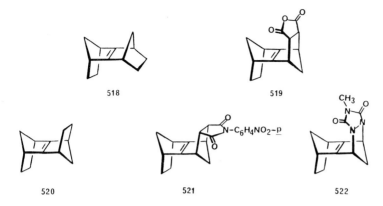

518   519

520   521   522

X-ray crystal analysis of several derivatives of *syn*-sesquinorbornene (518) has provided evidence that this ring system experiences large deviations from planarity about its central double bond (12—22°) [228—231]. To a lesser extent, structurally simpler norbornenes likewise exhibit equilibrium nonplanar character [232—234]. In contrast, most [228, 230, 231] though not all [235—237] *anti*-sesquinorbornenes possess an essentially flat π bond. The fact that 519 contains a planar double bond whereas that in 520—522 is puckered supports the notion that the molecular structures are fundamentally different.

The pair of tropone-isodicyclopentadiene adducts 523 and 524 offer a complementary set of structural features [238—240]. Both compounds share in common a butadiene bridge that is necessarily affixed diaxially to a cyclohexanone chair. Yet, the interplanar angle in 523 exhibits a significant departure from planarity, while that in 524 is essentially flat. Compounds 525 and 526 can be regarded as structural analogues of 523 [238, 239, 241].

168.6°   178.9°

523   524

170.7°   168.2°

525   526

Metal-assisted dimerization of norbornadiene provides the cage compound 527 whose structural features have now been determined crystallographically [242]. The crystal and molecular structure of [4]peristylanedione (528) has also been established by X-ray diffraction techniques. The molecule shows almost perfect C_{2v} symmetry, the four-membered ring and both five-membered rings

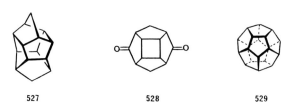

527   528   529

bearing the ketone groups are almost planar and the other two cyclopentane rings adopt half-chair conformations [243]. The long-awaited X-ray structure determination of dodecahedrane (529) has been achieved recently [244]. The geometry of this hydrocarbon does not deviate significantly from I_h symmetry and the exterior C-C-C bond angles conform nicely to the value expected for perfect dodecahedral symmetry (108°). Taking account of the van der Waals radius of carbon, the trans-cavity diameter in 529 is only 0.91—0.93 Å.

B Photoelectron Spectra

The photoelectron spetra of *syn*- (518) and *anti*-sesquinorbornene (520) show the ionization potentials to be 8.12 and 7.90 eV, respectively [245]. Proper comparison of these data with those of model systems indicates that both isomers are easier to ionize than expected. This phenomenon has been attributed to σ–π closed shell-closed shell repulsions between the ethano and methano bridges and the π bond.

Using photoelectron spectroscopy, Gleiter and coworkers have established that progression through the series 530 → 531 → 532 is accompanied by a strong increase in the level of interaction between the π-fragments [246, 247).

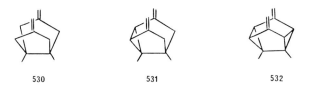

530   531   532

The He(I) photoelectron spectrum of 533 has been recorded and the first four bands assigned to ionization events from π-orbitals related to the perimeter of this [10]annulene [248].

533   534

74

C Molecular Orbital Calculations

The thermal rearrangement of 533 to 534 has been evaluated by MINDO/3 calculations [248]. The observation was made that the system is almost flat in the transition state, with the methyl group bridging carbons 7 b and 2 a. The stabilizing effect of the benzene ring that is formed in the transition state is estimated to be 15 kcal/mol^{-1}.

Molecular geometries have been calculated by the MNDO method for the cation and anion radicals derived from [3.3.3]propellane, as well as for the neutral triquinane and its dication [249].

No less than four research groups have given attention to the molecular and electronic features of the ground and/or excited stets of *syn*- (518) and *anti*-sesquinorbornene (520). Gleiter and Spanget-Larsen reported extended Hückel calculations that reproduce the magnitude of the pyramidalization in 518 and concluded that distortion occurs to minimize hyperconjugation between the π orbitals and the "ribbon" orbitals associated with monomethylene bridges [250]. Molecular mechanics calculations led Jorgensen to a very similar conclusion [251]. In contrast, Houk and his coworkers attribute the pyramidalization observed in the syn species to torsional effects [252]. Also, STO–3G and 3—21G calculations indicate that the triplet state of 518 is much more severely distorted in the opposite direction. In their hands, MM2 predicted 520 to possess a much more easily deformed π bond. Johnson has also performed empirical force field calculations on 535 amd 536 [253]. The ground state structure for 535 was found to have exact C_{2v} symmetry and to be bent 7° about the double bond. The smaller analogue 536 was determined to have a 35° fold about the double bond. Dynamic bending about this fulcrum in either compound would correspond to a potentially observable topomerization process.

535   536

D Miscellaneous

The heat of formation of perhydrotriquinacene has been obtained from that of adamantane; the enthalpy of isomerization was computed by ab initio SCF methods with correction for zero-point and thermal effects [254].

It now appears that the degree of circular polarization in luminescence can be utilized as a tool for the study of intramolecular excitation energy transfer. *trans*-Bicyclo[3.3.0]octane-3,7-dione (537) has served as a useful test probe of the

537

phenomenon since its two remote carbonyl chromophores are of opposite chirality and the substrate in its ground state is thus a meso structure [255].

^{13}C NMR spectral assignments to cedrol, 6-isocedrol, and α-cedrene have been corroborated using double quantum coherence measurements in tandem with long-range heteronuclear ^{13}C/^{1}H spin-spin coupling constants [256].

V Molecules of Theoretical Interest

A Pentalene

The effects of electron-donating substituents on the bond alternations in pentalene have been evaluated by using the Pariser-Parr-Pople-type SCF MO CI method in conjunction with a variable bond-length technique [257]. A valence bond treatment of aromaticity, first introduced by Mulder and Oosterhoff, has now been applied to pentalene [258]. MNDO, UMDDO, and MNDO/CI calculations have also been reported for pentalene [259].

In agreement with predictions based on MNDO calculations, a crystal structure determination of the dilithium pentalenide-dimethoxyethane complex shows the metal atoms to bridge opposite faces of the 10π aromatic system as in 538. This arrangement is thought to be favored because each carbon atom has a lithium neighbor, the lithium cations communicate electrostatically, and dipolar repulsion does not occur [260].

538

When the TMEDA solvate of 538 in dry tetrahydrofuran solution is irradiated at $-48\,°C$, the pentalene radical anion is produced. Its calculated structure is that obtained by removal of one lithium atom from 538 [261].

In an interesting series of transformations, azulene 539 has been transformed readily into the substitued pentalene 540 [262].

539

540

77

B Semibullvalenes

The Cope rearrangement in semibullvalenes and its perturbation by framework substitution continues to be investigated intensively. In one of the simpler systems recently studied ($541 \rightleftarrows 541'$), the exchange bonds the deuterium atom either to an sp^2-hybridized carbon or to a cyclopropane ring. The latter valence isomer is favored ($K_{25°C} = 1.11$) [263]. The trideuteriomethyl semibullvalenes 542a and 542b have also been studied and shown to prefer positioning the CD_3 substituent at the allyl site [264].

541   541′

542a, R= H
  b, R= COOCH₃

Dynamic ^{13}C NMR methods have been utilized to measure the activation barriers to Cope rearrangement in 543—545 [265]. The activation parameters, compiled in Table 1, in particular the $\triangle G^{\ddagger}$ values, reveal that all three isomerize more slowly than semibullvalene itself. Indeed, there is no need to invoke a biradicaloid transition state.

543   544   545

Table 1. Activation parameters for the Cope rearrangement of 543—545

	E_A [kcal/mol]	$\Delta H^{\ddagger}$ 25 °C [kcal/mol]	$\Delta G^{\ddagger}$ 25 °C [kcal/mol]	$\Delta S^{\ddagger}$ [e.u.]
	5.1	4.8	5.5	−5.4
543	6.7 ± 0.2	6.1 ± 0.2	6.1 ± 0.1	−0.2 ± 0.3
544	4.7 ± 0.5	4.1 ± 0.5	5.5 ± 0.1	−4.9 ± 5
545	6.2 ± 0.2	5.6 ± 0.2	6.0 ± 0.1	−1.2 ± 3

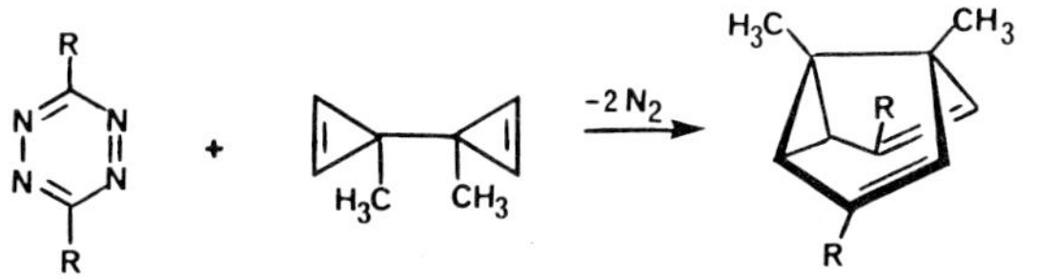

R=COOCH₃   546

The ordinary behavior of 545 holds particular importance since this semibullvalene is known from X-ray data to exhibit a rather long C-2/C-8 cyclopropane bond and a short nonbonding distance across the C-4/C-6 gap [266]. Consequently, the latter effects appear to be specific to the crystal lattice and should not be construed as evidence for a bishomoconjugated ground state. The same rationale applies to diester 546 which can be prepared in a one-step synthesis as shown [267].

The synthesis of 545 and its 2,6-dibromo (547) and 4-bromo derivatives (548) has been achieved by an alternative protocol (Scheme XXI) [267]. As expected,

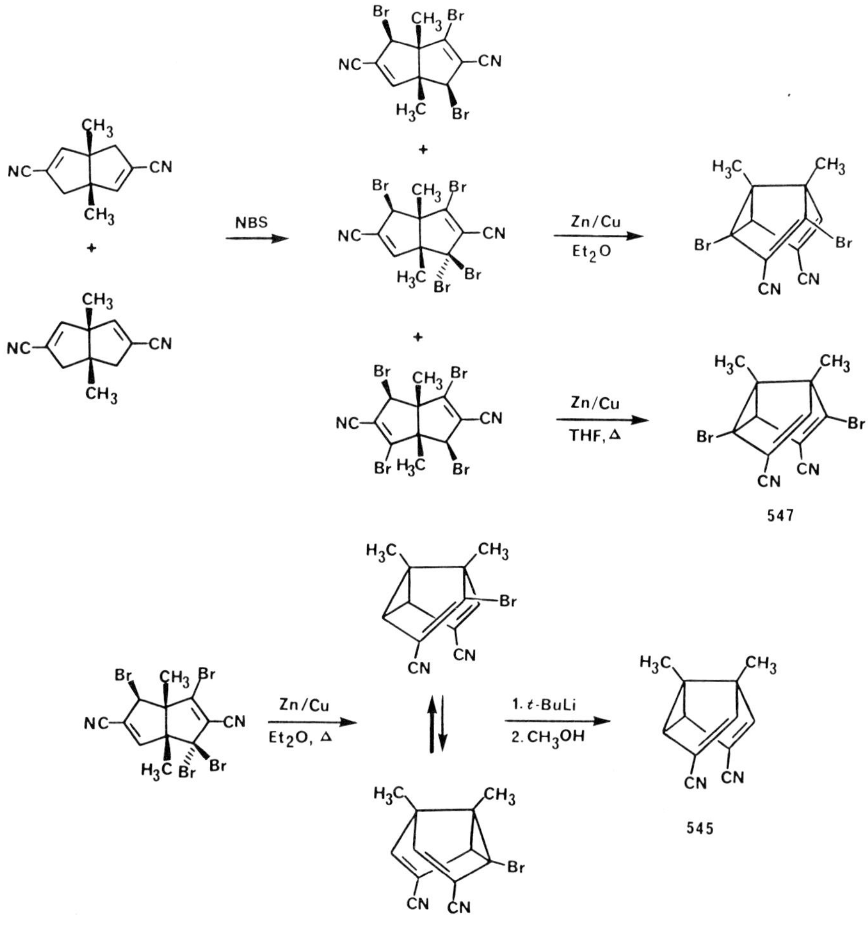

Scheme XXI

the presence of bromine atoms slows down the degenerate Cope rearrangement relative to 545. More importantly, cyano groups in the 3 and 7 positions raise the Cope barrier in striking contrast to the effect caused by cyano groups at C-2 and C-6 [268]. X-ray crystallographic data suggest that distortions in ground state geometries may correlate with the observed barriers to Cope rearrangement, especially when bromine substituents are present [269].

2,6-Dicyano-1,5-dimethylsemibullvalene (549) was prepared by zinc iodide-catalyzed addition of trimethylsilyl cyanide to bicyclo[3.3.0]octane-2,6-dione [270, 271], conversion to its unsaturated dinitrile with POCl$_3$ in pyridine, allylic bromination with N-bromosuccinimide, and reductive cyclization with zinc-copper couple (Scheme XXII) [268]. Dinitrile 549 exhibits the lowest of all

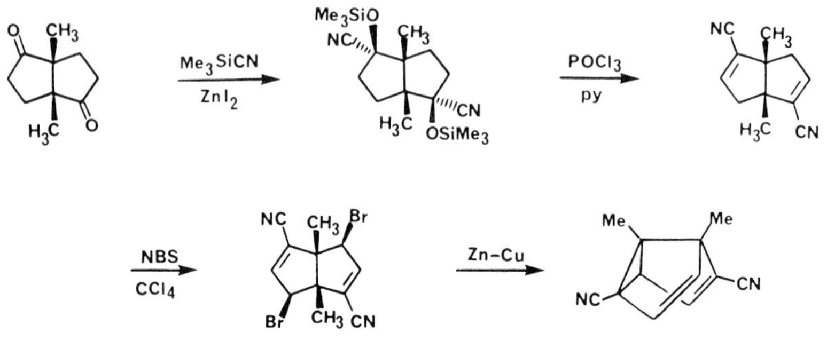

Scheme XXII

previously determined Cope activation barriers ($\triangle G^{+}_{115}$ = 13.0 ± 0.5 kJ/mol) [272]. The temperature dependence of its ultraviolet spectrum has prompted Quast to conclude that an unstable species isomeric with 549 exists in rapid equilibrium and that this species is the delocalized, bishomoaromatic structure 550. A new MNDO theoretical treatment of the rearrangements of these semibullvalene derivatives has been published [273].

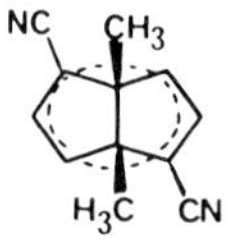

550

An ingenious isoxazole route to cyano-substituted semibullvalenes has been described by Askani and Littman [274]. They have taken advantage of the ease with which carboethoxyformonitrile oxide cycloadds to enamines 551 and 552 and subsequent Cope elimination of the nitrogen substituents (Scheme XXIII).

The pyrolysis of semibullvalene (T > 270 °C, 1 torr) results in the establishment of a reversible equilibrium with cyclooctatetraene, as demonstrated by approach from both sides. The equilibrium value of semibullvalene is 2–4% [275]. The thermal isomerization of 549 to 553 has also been reported [268].

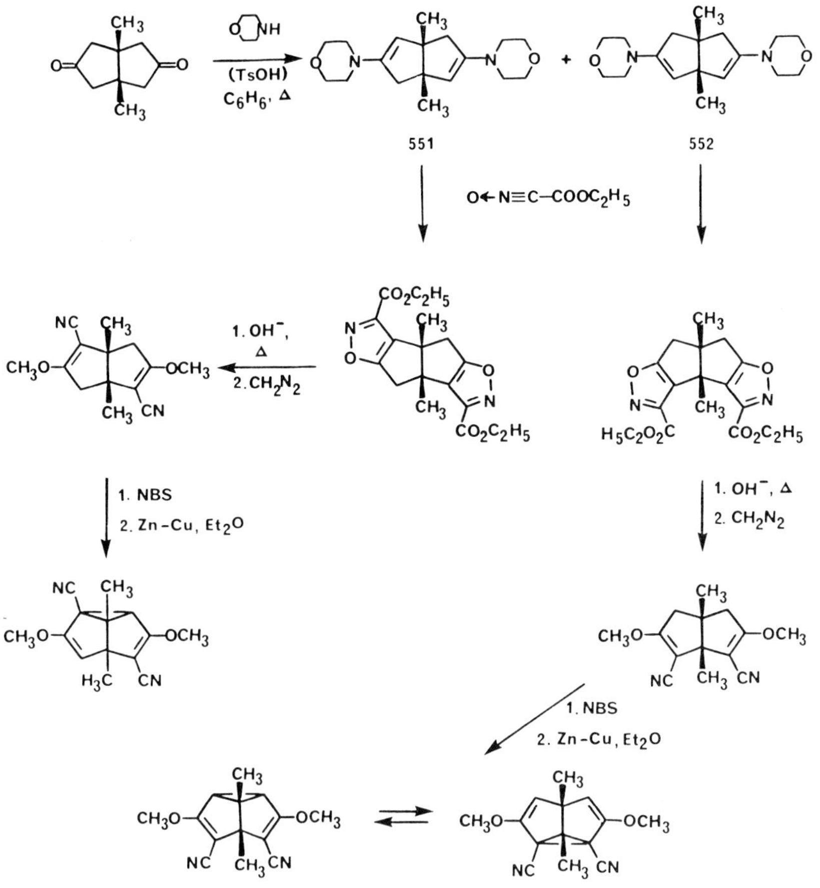

Scheme XXIII

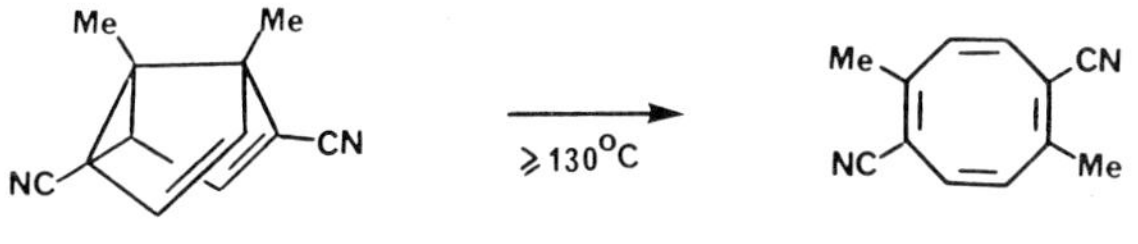

The two previously undefined dilithium semibullvalenides have now been identified as the C_{2h} and D_2 diastereoisomers of bis-(bicyclo[3.3.0]octa-3,7-diene-2,6-diyl)tetralithium [276].

C *Syn/anti*-Sesquinorbornenes

Kopecky and Miller have developed the route to 520 illustrated in Scheme XXIV [277]. The *syn*- and *anti*-isomers of sesquinorbornatriene (554 and 555) have also recently yielded to synthesis (Scheme XXV) [278–280]. Cycloaddition of (*Z*)-1,2-bis-(phenylsulfonyl)ethylene to tricyclo[5.2.1.0^{2,6}]deca-2,5,8-triene proceeds with at least 95% below-plane stereoselectivity to give the *syn* [4 + 2] adduct.

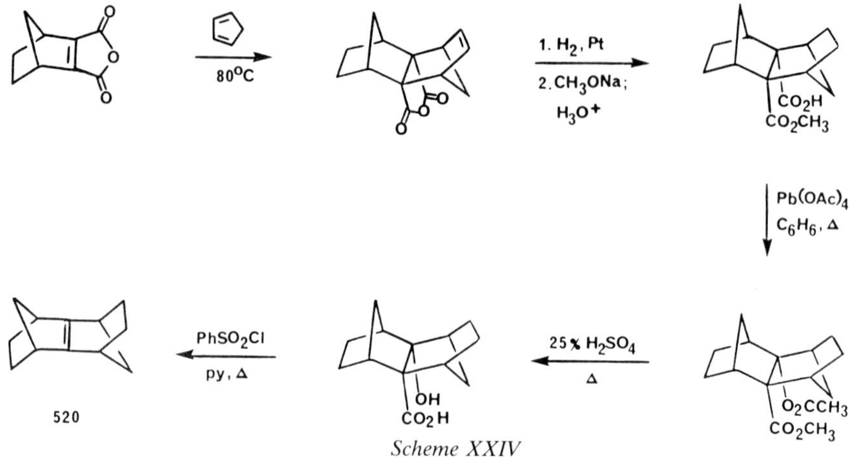

520

Scheme XXIV

Mild reductive desulfonylation of this product leads to 554. By comparison, the Diels-Alder reaction of tricyclo[5.2.1.$0^{2,6}$]deca-2,5,8-triene with (*E*)-1,2-bis-(phenylsulfonyl)ethylene is sterically controlled and proceeds with predominant above-plane dienophile capture. Although subsequent conversion to 555 revealed the hydrocarbon to be sensitive to air oxidation, the reactivity level is appreciably less than that of its syn counterpart.

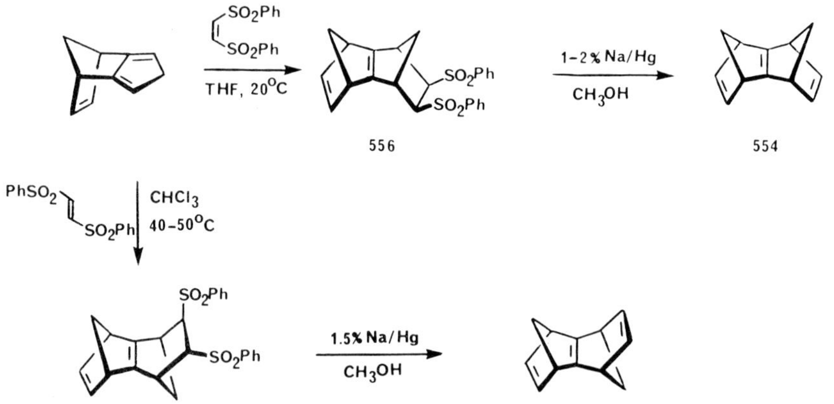

556 554 555

Scheme XXV

Sensitized irradiation ($PhCOCH_3$) of 556 successfully produces the air-stable quadricyclane disulfone 557. Treatment of this product with 1–2% sodium amalgam afforded the structurally novel substance 558 [278].

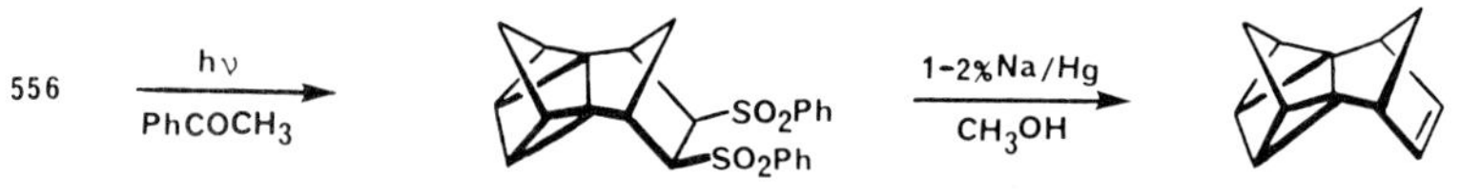

557 558

The cycloaddition of cyclopentadiene to 559 proceeds with bonding to the unsubstituted side affording predominantly 560 and isomers [279].

559   560   + isomers

An MM2 model to calculate the relative energies of stereoisomeric transition states of Diels-Alder reactions of isodicyclopentadienes and substituted derivatives has been developed by Brown and Houk [281]. Their conclusions are that torsional factors influence the preference for attack on the bottom face and that steric effects on either the dienophile or the isodicyclopentadiene can override this preferred stereoselectivity. A parallelism has been noted between dienophilic capture and singlet oxygenation of this system [282]. On the other hand, triazolinediones have proven to be "maverick" dienophiles [283]. Thus, with isodicyclopentadiene, attack occurs entirely above-plane to give 561; its dehydro counterpart gives rise uniquely to 562.

561

562

10-Isopropylideneisodicyclopentadiene (563) has been synthesized and its stereoselective behavior during Diels-Alder addition examined in detail. Less control of π-face selectivity was seen relative to the control exhibited by the parent isodicyclopentadiene [284]. The three 7-spirocyclopropyl-substituted analogues 564—566 were also prepared and comparably studied. Whereas diene 564 exhibits no strong predilection for [4 + 2] cycloaddition to either face, furan 566 enters into Diels-Alder reaction totally by above-plane bonding. This contrasting behavior is believed to have an electronic origin. On the other hand, the response of congested diene 565 is believed to be controlled entirely by intermolecular steric interactions [285].

563   564   565   566

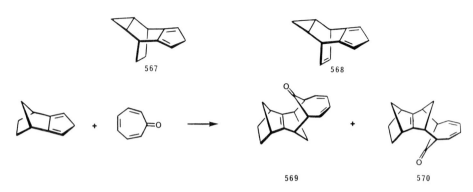

The π-facial stereoselectivities exhibited by 567 and 568 have been deter-mined [286]. Cycloaddition of tropone to isodicyclopentadiene occurs predomi-nantly by a [6 + 4] bonding process with a marked above-plane selectivity. The two most prevalent ketones 569 and 570 were formed in a 71:15 ratio [287–289]. The presence of a 2-substituent on the tropone is sufficient to render this process inoperative in favor of capture of the [1,5] sigmatropic diene isomers.

The 1,3-diphenyl-2-oxyallyl dipolar ion, generated by several reductive methods and therefore complexed to metal ions of various type, has been added in [3 + 4] fashion to isodicyclopentadiene. Five (571—575) of the six possible adducts were formed. In contrast, the tetramethyloxyallyl cation gives rise to a high preponderance of 576 [287, 290].

The course of various Diels-Alder cycloadditions to 577—579 has been examined with a view to gaining insight into possible silatropic migrations within these systems [291]. Bartlett and Wu have studied the reactions of tetracy-anoethylene [292], maleic anhydride [293], and dimethyl acetylenedicarboxylate

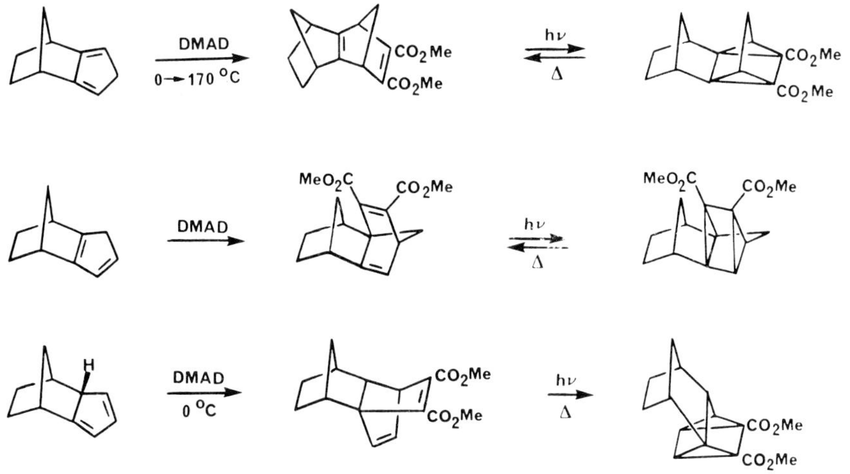

DMAD
0→170 °C

CO₂Me
CO₂Me

hν
Δ

CO₂Me
CO₂Me

DMAD

MeO₂C CO₂Me

hν
Δ

MeO₂C CO₂Me

DMAD
0 °C

CO₂Me
CO₂Me

hν
Δ

CO₂Me
CO₂Me

Scheme XXVI

[293] with the three isodicyclopentadiene isomers [294]. A typical set of results is given in Scheme XXVI.

The rates of a number of concerted, free radical, and ionic additions to the double bond of *syn*- and of *anti*-sesquinorbornenes have been determined. Equilibria, where measurable, are more favorable to addition with the anti than with the syn isomer. The rate ratio k_{syn}/k_{anti} varies from less than 0.05 for reversible, ionic additions to as high as 1–7 for free-radical and concerted reactions [295, 296].

The double bond in 580 is from 1.3 to 13.7 times more reactive than that of 581 [297]. The oxabenzo-*syn*-sesquinorbornene is also rearranged either by ultraviolet light or acid to 582 [298].

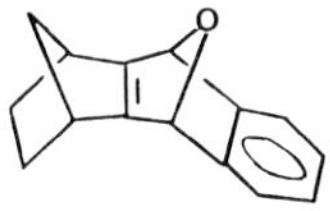

OH

580 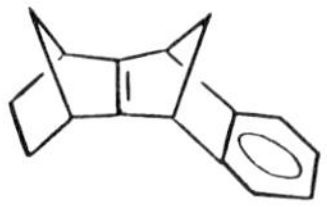   581 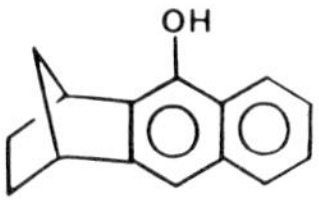   582

When photosensitized in acetone solution, *syn*-sesquinorbornene experiences hydrogenation mainly from its endo face [299]. This remarkable result is due to deformation of the central π bond in the exo direction in the triplet excited state of 518 [252].

D Triquinacenes

A new synthetic route to triquinacene (587) has been devised having as its basis the deployment of the Weiss-Cook reaction [300]. Compound 583 results from the condensation of glyoxal with dimethyl acetonedicarboxylate (Scheme XXVII). High-yield monoalkylation of 584, aldol cyclization of 585, and HMPA-promoted dehydration of triol 586 comprise the other key steps.

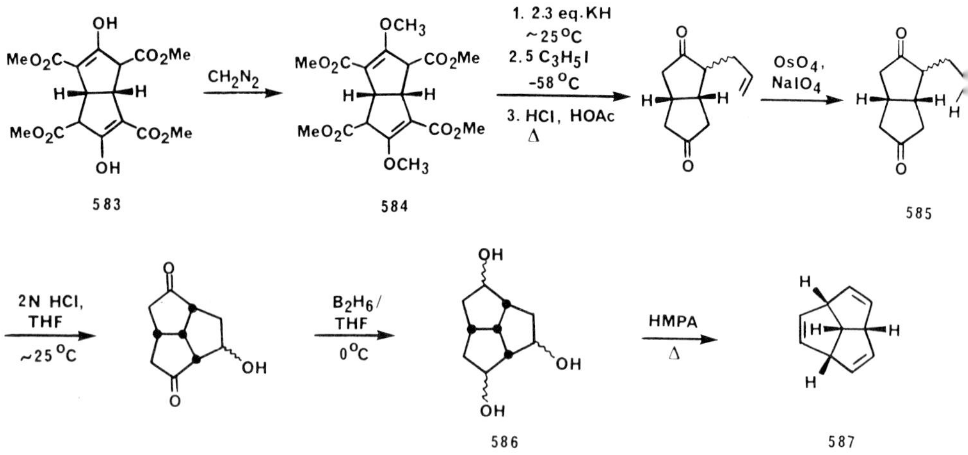

583   584   585

586   587

Scheme XXVII

Triketone 591, a useful precursor to triquinacene, has been prepared by application of the intramolecular Pauson-Khand reaction. Opening of lactol 588 with ethynylmagnesium bromide, suitable blocking of the two hydroxyl groups, and silylation of the terminal alkyne furnishes 589. Closure to 590 proceeds in 82% yield (Scheme XXVIII) [301].

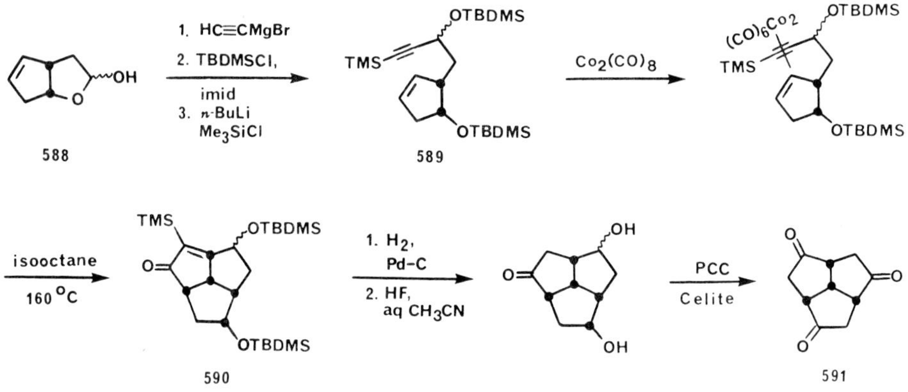

588   589

590   591

Scheme XXVIII

Whereas dimesylate 592 in dichloromethane solution readily undergoes twofold elimination when slurried with activated alumina at room temperature to give triquinacene, exposure to potassium *tert*-butoxide in anhydrous dimethyl sulfoxide instead leads predominantly to isotriquinacene (593) [302].

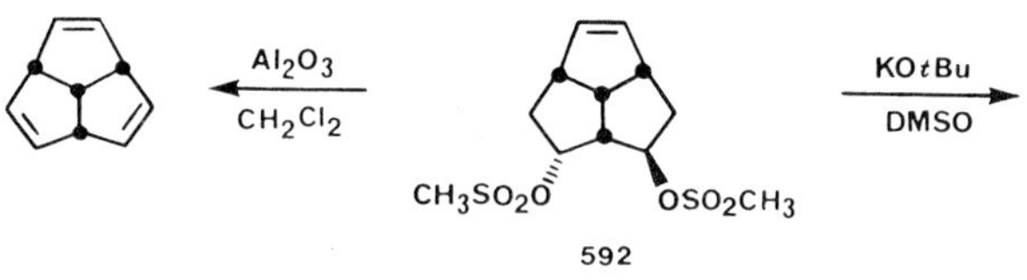

592   593

1-Halo- (594) and 1,4-dihalotriquinacenes (595) readily react with secondary amines and alkyllithium compounds to yield related bridgehead olefinic and

double bridgehead olefinic derivatives of triquinacene [303]. These developments and the stepwise conversion of 595 to 596 and 597 are indicative that such systems differentiate between hard and soft nucleophiles. With hard electrophiles, only bridgehead substitution products are formed; soft nucleophiles yield 3-isotriquinacene derivatives instead. In contrast, the trihalo compounds

yield bridgehead substitution products, e.g. 598, with essentially all types of nucleophiles [304]. Another interesting reaction occurs when tetraenes such as 599 are reduced with sodium metal. Tris(dialkylamino)triquinacenes (600) are formed, presumably by a reduction/disproportionation mechanism [304].

Diamine 599 is readily transformed into its tricarbonyliron complex 601, which upon two-electron reduction gives the novel tricarbonylferrate 602. 1,4-Dibromo-, 1,4,7-tribromo-, and 1,4,7-trichlorotriquinacene react with $Fe_2(CO)_9$ in tetrahydrofuran to yield the (dihydroacepentalene)hexacarbonyldiiron complexes 603 and 604 [305].

E Tricyclic [10]Annulenes

Diester 605, formed by acid-catalyzed elimination of methanol from the [8 + 2]-cycloadduct of 3-methoxy-3a-methyl-3aH-indene and dimethyl acetylenedicarboxylate, possesses a 10π-electron aromatic periphery that sustains a ring current. With cupric nitrate and acetic anhydride, 605 gives rise to a mixture of mononitro substitution products [306]. Conversion to dialdehyde 606 and decarbonylation with tris(triphenylphosphine)rhodium(I) chloride provides a route to the unsubstituted system 607 [307].

A better overall route to 607 begins with ketone 608, Shapiro degradation of which delivers the acid-sensitive 609 [307]. Both of these approaches have been extended, the first to gain access to the ethyl homologue (610) and the second to prepare the isopropyl derivative (611, Scheme XXIX) [308].

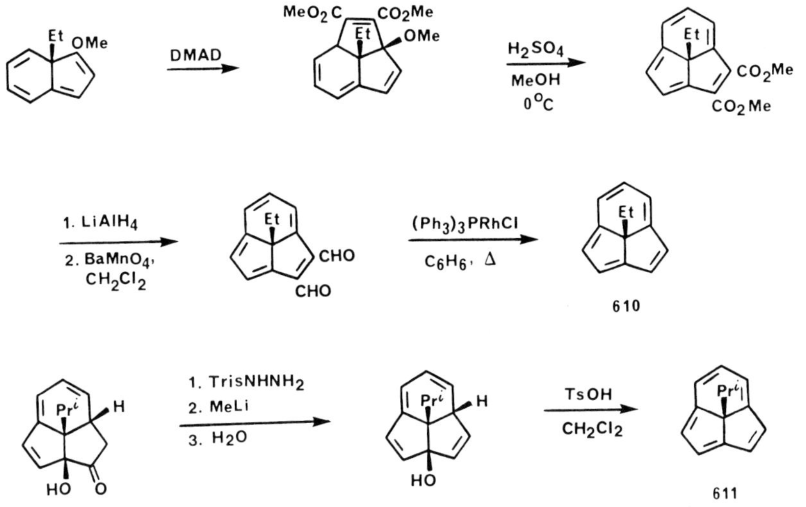

Scheme XXIX

The tricyclic [10]annulene containing a central benzyl group has been prepared. Reductive alkylation of 7-methoxy-1-indanone to give dienone 612, introduction of a third double bond, and condensation of the resulting trienone with methoxyvinyllithium afforded alcohol 613 (Scheme XXX) [309]. Methylation and acid hydrolysis gave diketone 614, whose aldol condensation was followed by a series of conventional steps.

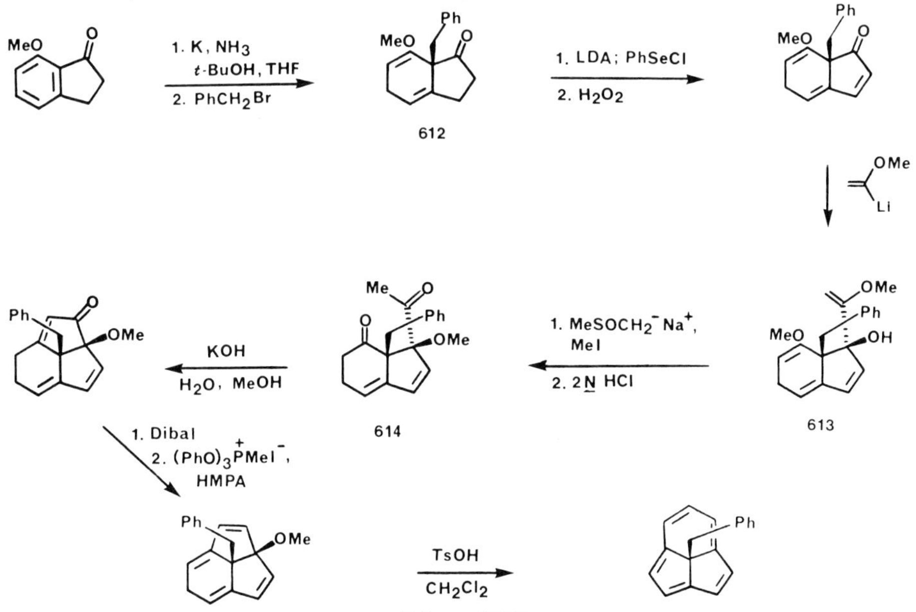

Scheme XXX

The effect of benzo-fusion on the tricyclic [10] annulene framework has been assessed by the synthesis of 615 and 616. Accurate bond lengths in 615 have been

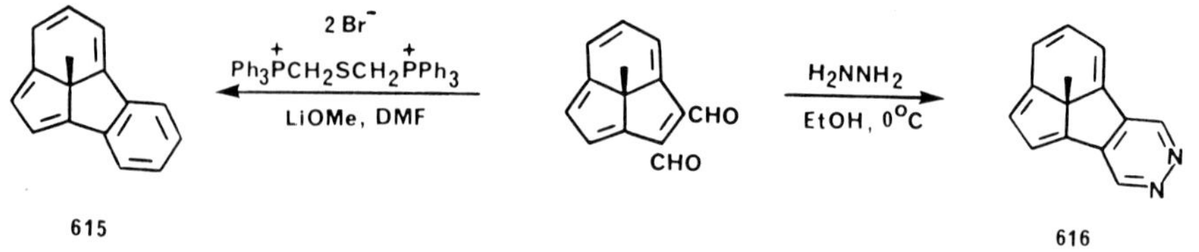

615

616

determined by X-ray crystallography. The substance retains approximately two-thirds of the ring current of 607 [310].

The 2- (621) and 5-hydroxy (626) derivatives of 607 have been synthesized and their properties compared. O-Methylation of 617 followed by cycloaddition with 2-chloroacryloyl chloride furnishes 618, a convenient precursor of 619 [311]. O-Silylation of this intermediate occurs with loss of methanol to give 620, hydrolysis of which provides 621 ⇌ 622 (Scheme XXXI) [312]. The route to 626 begins with 623 and bears many similarities to earlier work. The demethylation of 624 proceeds upon addition of HBr and furnishes 625 [312, 313].

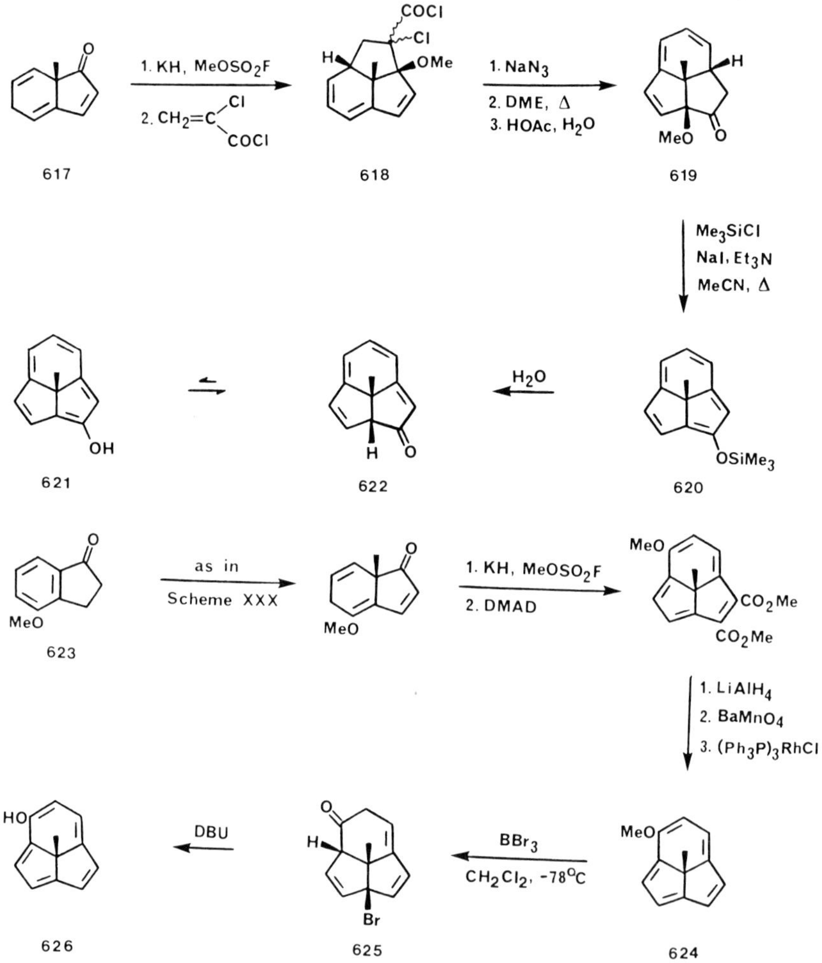

Scheme XXXI

Although the 2-hydroxy isomer exists entirely in its nonaromatic keto form (622), its lithium enolate does sustain a substantial diamagnetic ring current. In contrast, the 5-hydroxy isomer (626) exists entirely in the annulenol form [312, 313].

The conversion of 617 to 618 appears to hold some generality. Some additional examples are illustrated. The increased migratory aptitudes of substituents other than methyl at the ring junction does, however, limit the scope [314].

Although the annulene 607 does not undergo cycloaddition with TCNE, DMAD, or benzyne, it does give a 2:1 adduct with N-phenyltriazolinedione. The hydrocarbon is equally responsive to chlorosulfonyl isocyanate; in this instance the ring-expanded indenoazepine 627 is formed, subsequent hydrolysis of which furnishes 628 [315].

F Fenestranes and Related Molecules

Polycyclic hydrocarbons in which four rings are annulated in such a way that they share a common carbon atom are of interest for the study of the planarization of tetracoordinate carbon. Toward this end, Keese and his coworkers have examined the intramolecular meta-photocycloaddition of (RS, SR)-3-phenyl-6-hepten-2-ol (629) as a possible route to [5.5.5.5]fenestranes [316]. When photolyzed in degassed hexane, 629 afforded six products, five of which (630–634) could be obtained pure by chromatography. The relative ratio of these alcohols was determined to be 1:0.5:0.6:0.08:0.4. The regio- and diastereoselectivity observed can be understood in terms of a vinyl group that is highly mobile in a conformational sense.

On heating, 630 undergoes [1,5]-hydrogen migration to give 635. Treatment with perchloric acid leads instead to hydration and formation of 636. Oxidation of 636 affords 637.

A most direct method for arriving at all-cis-[5.5.5.5]fenestrane (639) involves reductive decarboxylation of lactone 638 [317].

Dicyclopentadiene serves as the starting material for a alternative route to 639. Conversion to dinitrile 640 can be achieved in several steps (Scheme XXXII). Under Thorpe-Ziegler conditions, 640 was readily transformed into β-enamino nitrile 641. Once keto lactone 642 had been obtained, it was treated with palladium and hydrogen at high temperature in the same manner as previously applied to 638. Again, 639 was isolated [318].

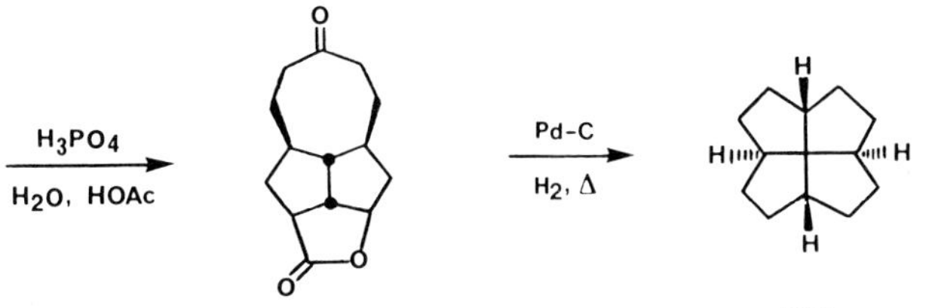

642 639

Scheme XXXII

The stage now appeared set for preparation of the more highly strained [5.5.5.5]-fenestrane stereoisomer 646. Following their earlier procedure, Luyten and Keese prepared dinitrile 643 and subsequently keto lactone 644 a. Photolysis of the potassium salt of trishydrazone 644 b gave instead of the desired cyclization product the olefin 645 or its regioisomer [319]. On the other hand, heating of 644 a with palladium and hydrogen gave 639 and not 646!

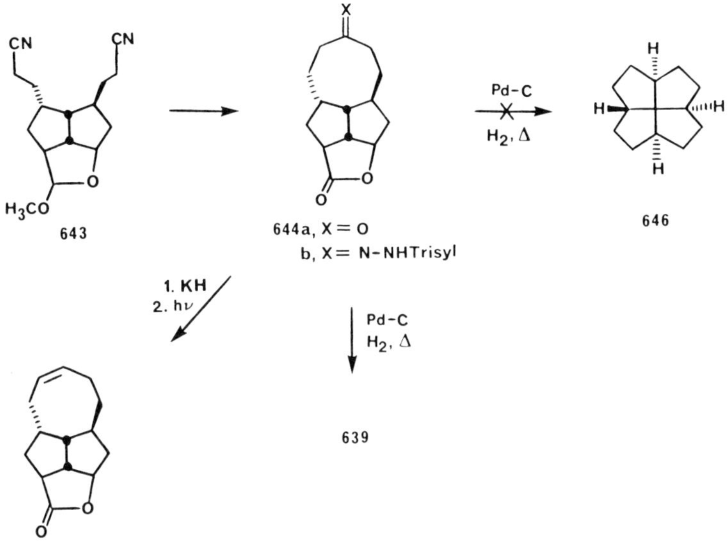

643
644a, X = O
b, X = N-NHTrisyl
646
639
645

At about the same time, the research group headed by Agosta noted that photolysis of dienone 647 furnished 648, which by way of diazoketone 649 was cyclized via carbene insertion into 650 (Scheme XXXIII) [320, 321]. Reductive removal of the carbonyl group in 650 and deketalization to provide 651 was followed by photochemical Wolff rearrangement of the derived α-diazoketone 652 in methanol. The [4.4.4.5]fenestrane esters 653 and 654 were thereby obtained. X-ray crystallographic analysis of the *p*-bromoanilide of 653 has provided important structural information.

The methyl group in 647 is present in order to control the regiochemistry of the [2 + 2] photocylization. It so happens that a chlorine atom can play the same role and deliver 655. A series of reactions parallel to that deployed earlier was utilized to arrive at the parent unsubstituted hydrocarbon 656 (Scheme XXXIV) [322].

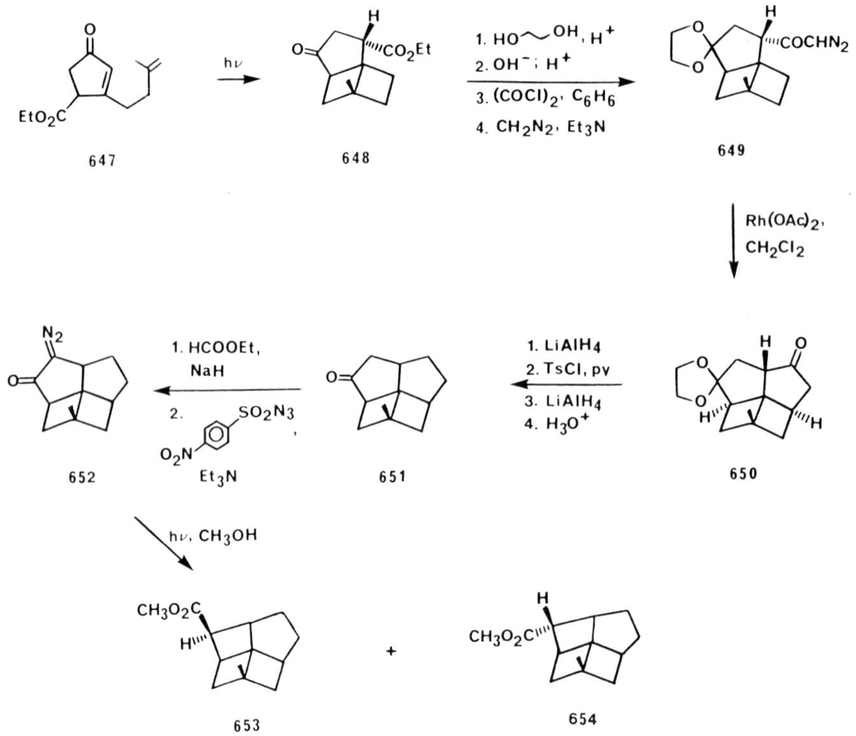

Scheme XXXIII

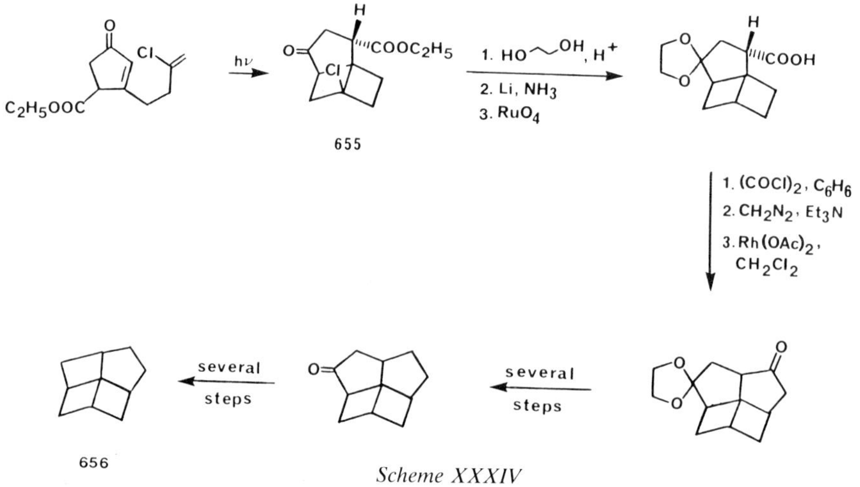

Scheme XXXIV

Somewhat related product control has been uncovered during the intramolecular arene-olefin photocyclization of 657. In this instance, 658 and 659 are obtained. The first of these has been transformed into the functionalized all-*cis*-[5.5.5.5]fenestrane 660 (Scheme XXXV) [323].

Contrary to the behavior of 644 b, the deprotonated form of trisylhydrazone 661 gives the bridged all-*cis*-fenestrane 662 on irradiation [324].

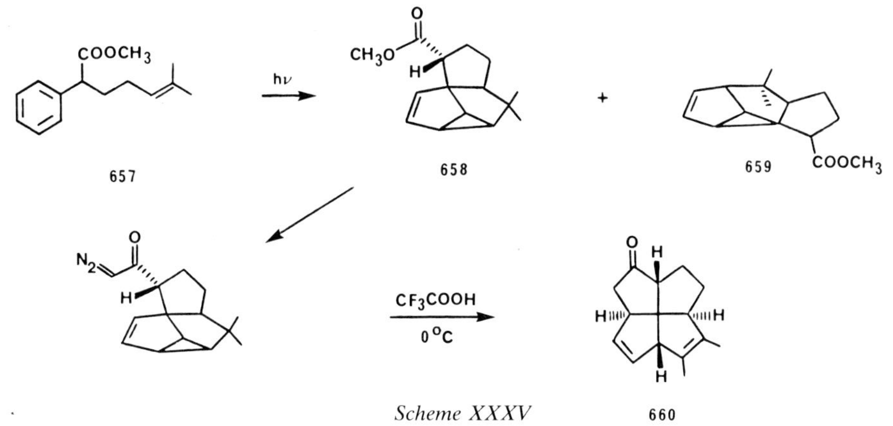
COOCH3
hν
CH3O
657
658
659
COOCH3
N2
CF3COOH
0 °C
660

Scheme XXXV

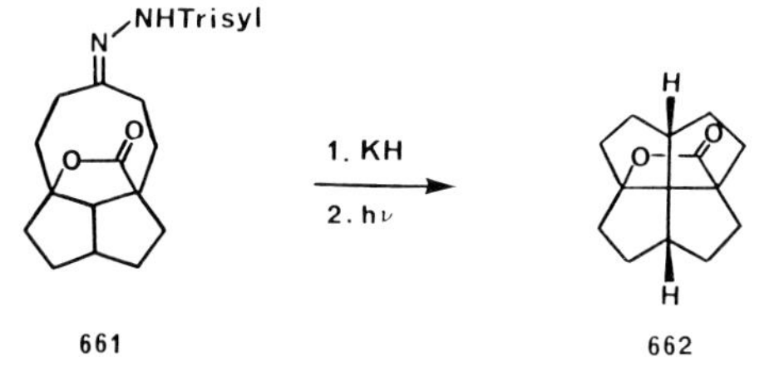
NHTrisyl
1. KH
2. hν
661
662

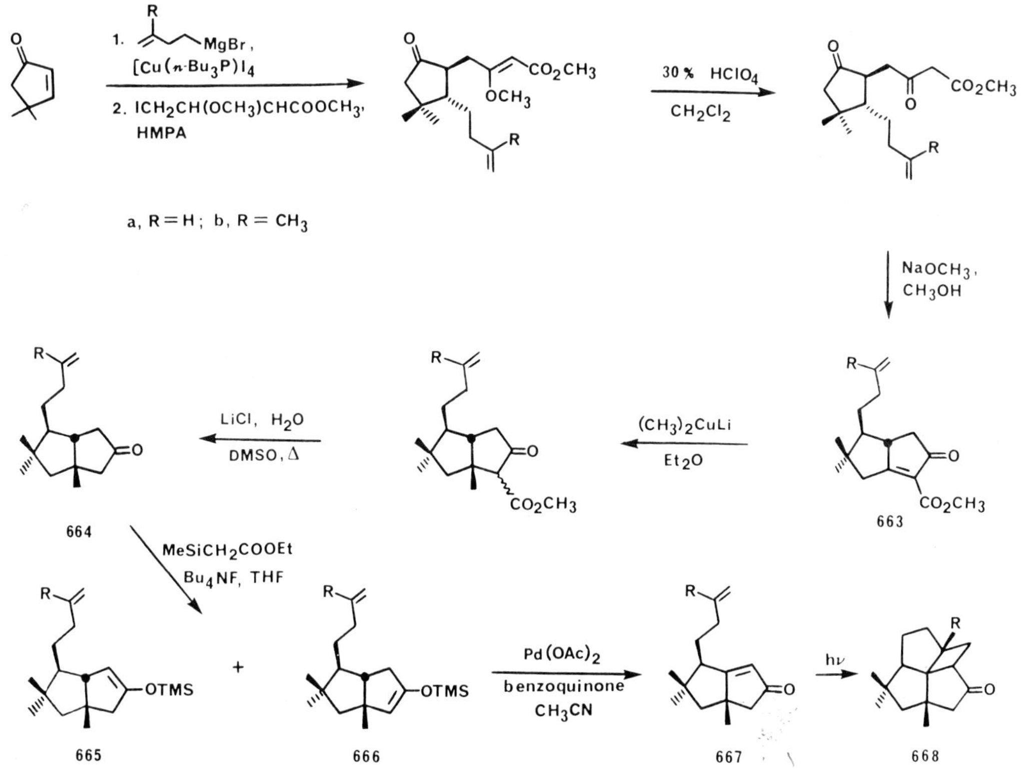
1. MgBr,
[Cu(n-Bu3P)I]4
2. ICH2CH(OCH3)CHCOOCH3,
HMPA
CO2CH3
OCH3
30% HClO4
CH2Cl2
CO2CH3
a, R=H; b, R=CH3
NaOCH3,
CH3OH
LiCl, H2O
DMSO, Δ
(CH3)2CuLi
Et2O
CO2CH3
664
663
MeSiCH2COOEt
Bu4NF, THF
OTMS
OTMS
Pd(OAc)2
benzoquinone
CH3CN
hν
665
666
667
668

Scheme XXXVI

The first example of a fenestrane which has cis substituents at opposing ring junctures has been synthesized via a high-temperature intramolecular photocycloaddition as the key step (Scheme XXXVI) [325]. Following assembly of keto ester 663 b, the ketone 664 b is obtained and transformed into the regioisomeric silyl enol ethers 665 b and 666 b. Oxidation of the latter gives rise to 667 b, which when irradiated at 110 °C in chlorobenzene afforded 668 b in 65% yield. The less strained congener 668 a has been produced in identical fashion; the dramatically lessened steric congestion was apparent in the much less stringent conditions necessary for formation of the cyclobutane ring (hv, hexane, 2 h).

The tetraunsaturated [5.5.5.5]fenestrane 673, dubbed staurane-2,5,8,11-tetraene, has been accessed by a Weiss-Cook condensation involving keto aldehyde 669 (Scheme XXXVII) [326, 327]. Osmylation of diketone 670 followed by Jones oxidation provided diacid 671. Heating this intermediate in the presence of naphthalenesulfonic acid (NSA) promoted > 70% closure to tetraketone 672. The most efficient method uncovered for introducing the four centers of unsaturation involved exhaustive reduction with diborane followed by heating in HMPA for 2 days.

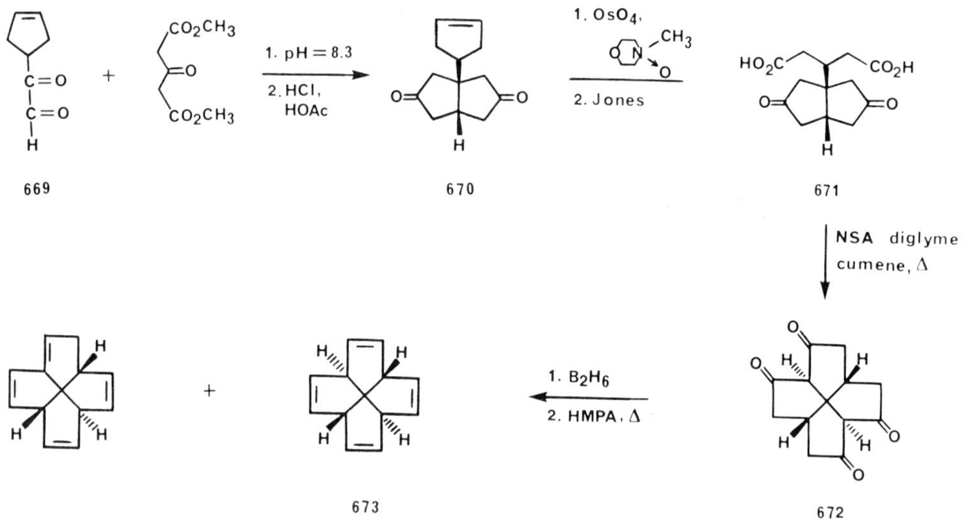

Scheme XXXVII

MINDO/3 calculations indicated that 674 should be nonplanar (D_2 symmetry) and show pronounced bond alternation in its ground state [328].

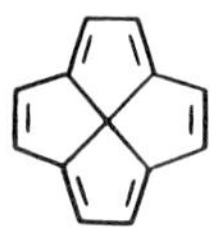

674

Similar methodology has been employed in the synthesis of the two structurally related polyquinanes 675 and 676 (Scheme XXXVIII) [327, 329, 330].

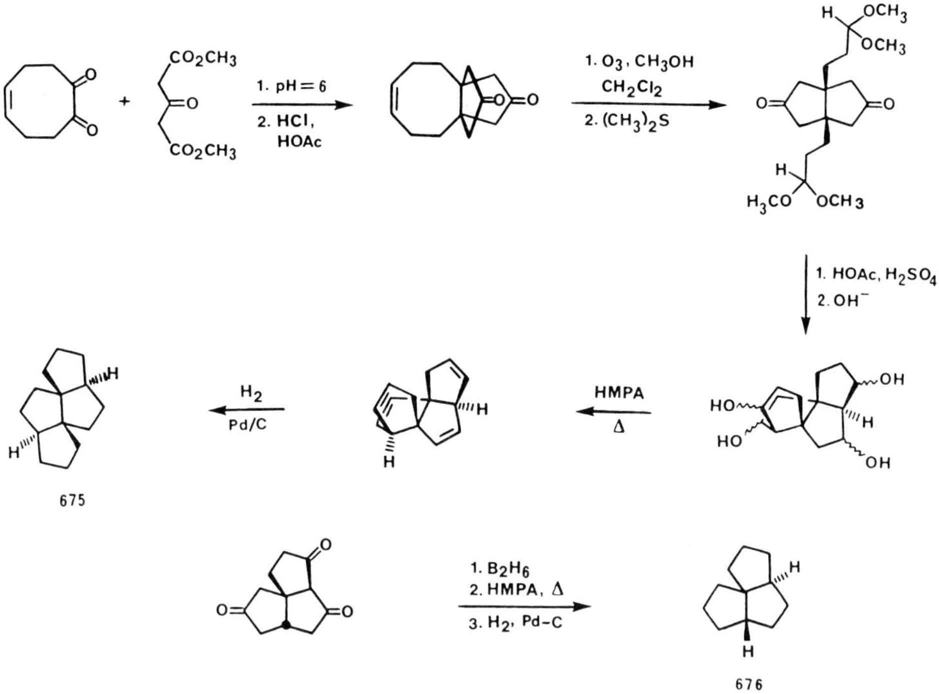

Scheme XXXVIII

G (D_3)-Trishomocubanes and Congeners

Optically active [3.1.1], [4.1.1], [5.1.1], and [6.1.1]triblattanes (678) have been prepared by diazomethane ring expansion of (—)-D_3-trishomocubanone (677) [331].

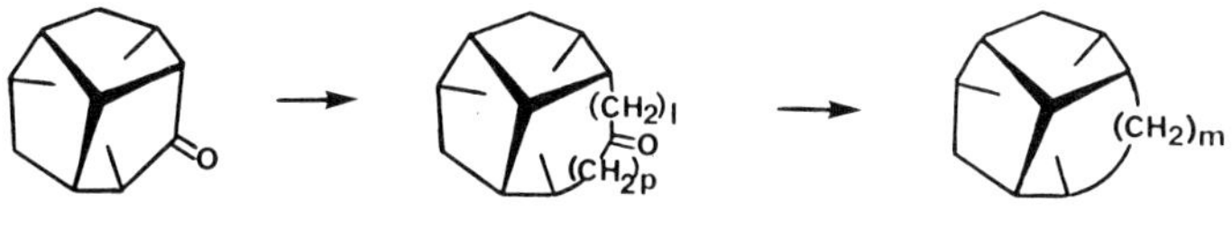

Ketone 679 was resolved into its dextrorotatory enantiomer by horse liver alcohol dehydrogenase-mediated reduction. Its reaction with diazomethane in the presence of boron trifluoride etherate followed by Wolff-Kishner reduction afforded a mixture of [m. 1]triblattanes (680) with m = 1–7. The hydrocarbons were individually isolated in a pure state by preparative VPC [332].

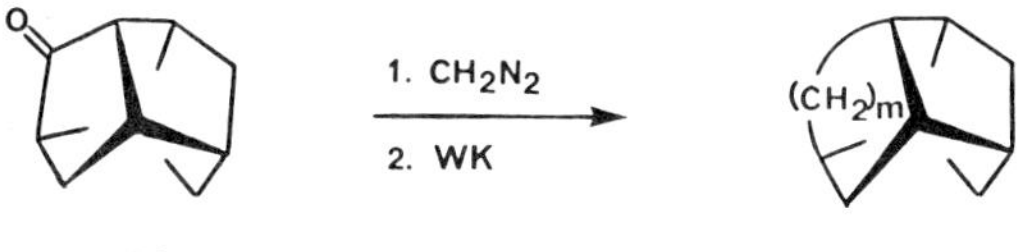

97

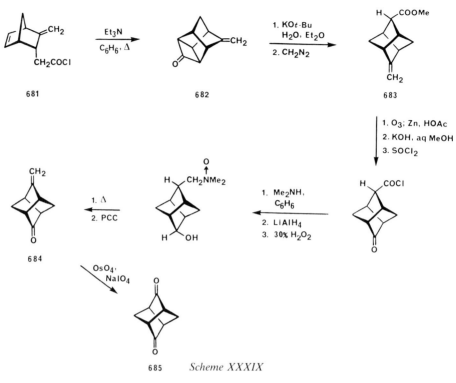

Scheme XXXIX

Heating acid chloride 681 with triethylamine yields ketone 682 via the ketene intermediate. Base-promoted cleavage of the four-membered ring proceeded regiospecifically to give 683 after esterification. Through the subsequent series of steps illustrated in Scheme XXXIX, 6-methylene-D_{2d}-dinoradamantan-2-one (684) was obtained. The action of OsO_4-$NaIO_4$ on 684 yields the dione 685 [333].

Fessner and Prinzbach have succeeded in preparing D_3-trishomocubanetrione (687) from 686 (Scheme XL) [334]. An efficient procedure for the optical resolution of *rac*-687 via the diasteromeric ketals 688 and 689 has also been described.

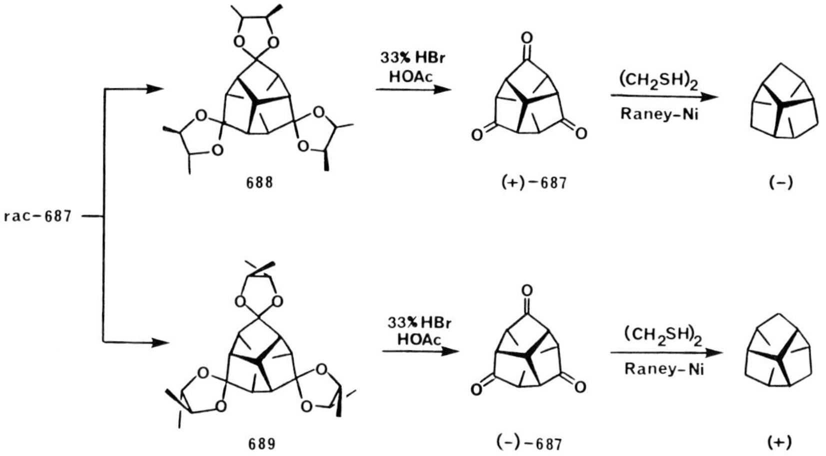

Scheme XL

The first optically active organic molecule having *T* symmetry has recently been synthesized in a manner where the absolute configuration is known. The levorotatory trishomocubaneacetic acid 690 was transformed by way of 691 into the acetylene 692. Activation of the triple bond was achieved by bromination of

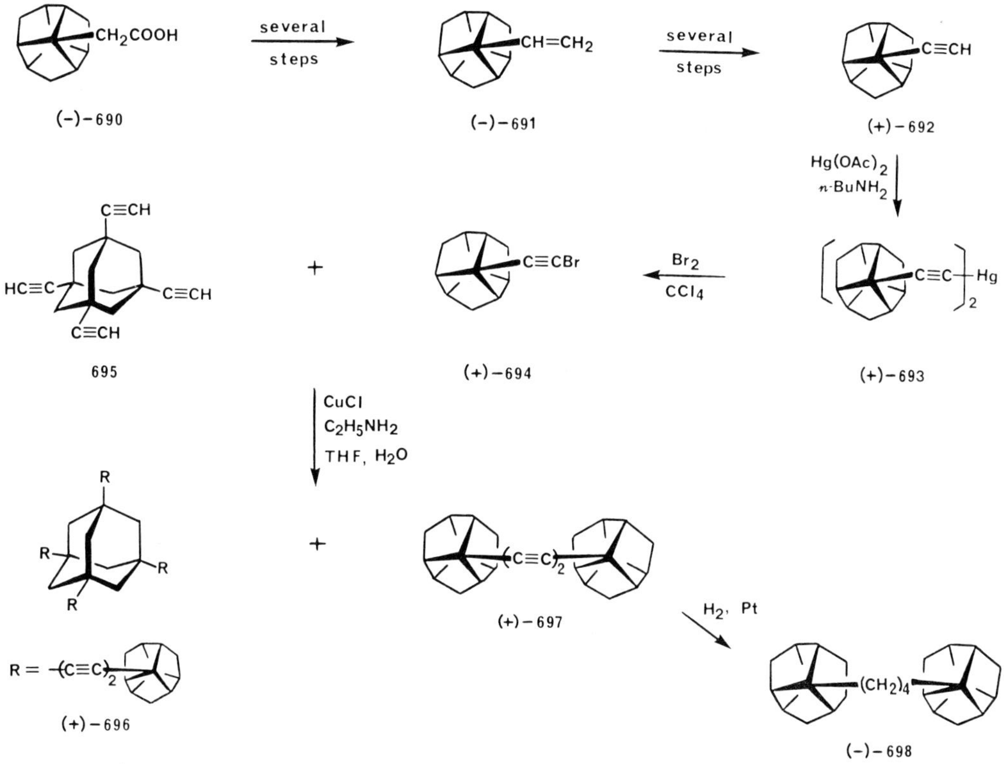

Scheme XLI

the mercury derivative 693 (Scheme XLI). Coupling of (+)—694 with the substituted adamantane 695 afforded a mixture of target molecule 696 and dimer 697. Hydrogenation of 697 gave 698 [335].

H Peristylanes

Carbene 701, generated either by thermolysis of 699 or photolysis of 700, undergoes ready γ C–H insertion to give methylene[3]peristylane (702) [336].

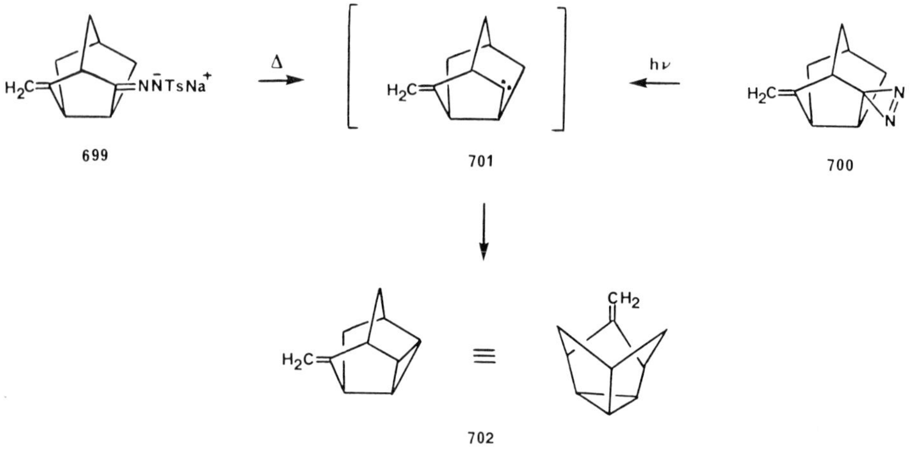

H2C=
=NN̄TsNa+
Δ
H2C=
hν
H2C=
N
N
699
701
700
H2C=
CH2
702

A short route to the capped [3]peristylanes 705–708 has been reported. Two-carbon chain extension of ketone 703, chemospecific reduction, epoxidation, and anionic cyclization forms the first additional framework C—C bond as in 704. Following tosylation, the cyclopropane ring in 705 is elaborated under anionic conditions (Scheme XLII). Of the remaining transformations, the direct decyanation to give 706 is particularly noteworthy [337].

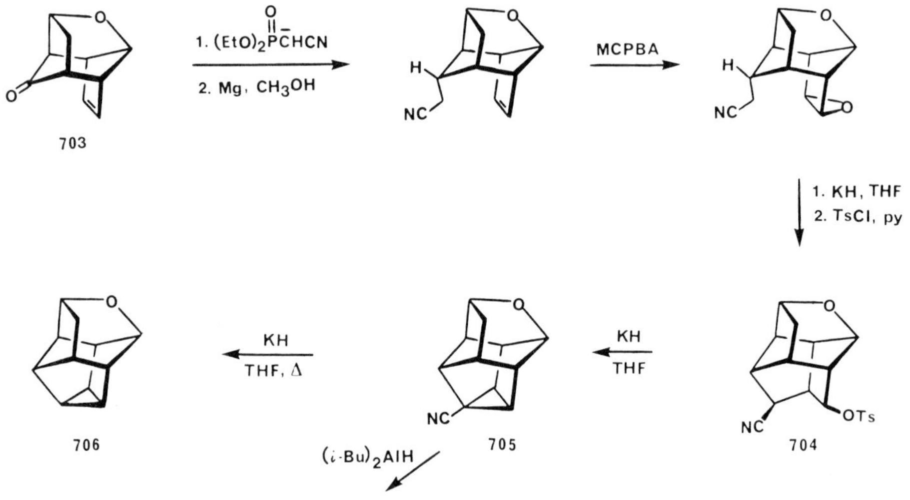

O
1. (EtO)2P(O)CHCN
2. Mg, CH3OH
H
NC
MCPBA
703
1. KH, THF
2. TsCl, py
KH
THF, Δ
KH
THF
NC
OTs
706
705
704
(i-Bu)2AlH

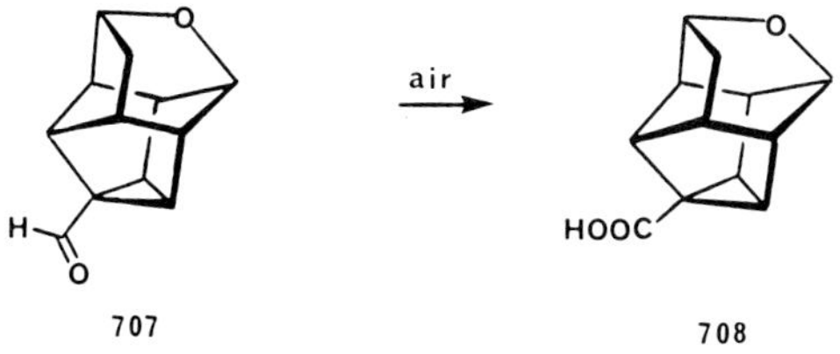

707 → 708

Scheme XLII

[4]Peristylane (711) has been prepared by *in situ* epoxidation of 709, the Diels-Alder adduct of (*p*-tolylsulfonyl)acetylene and tricyclo[5.2.1.0^{2,8}]deca-2,5,8-triene (Scheme XLIII) [338]. Arrival at the proper framework materialized upon [2 + 2] photocyclization of 710 and subsequent periodic acid cleavage in aqueous methanol. Once removal of the arenesulfonyl group was accomplished, the pair of carbonyl groups in 528 were reduced and tosylated prior to final reductive deoxygenation.

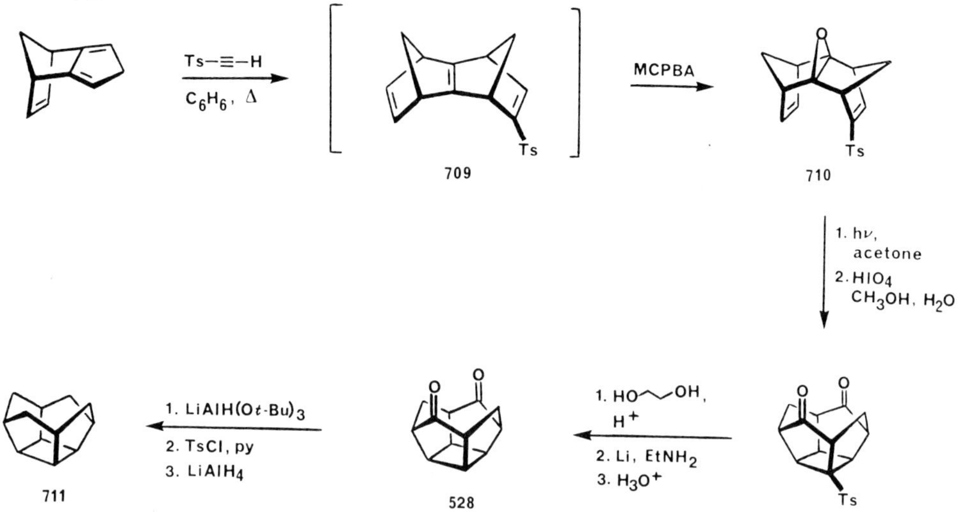

Scheme XLIII

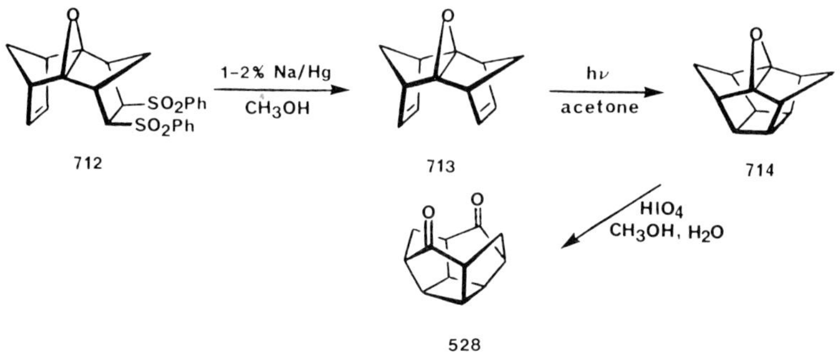

The readiness with which (*Z*)-1,2-bis(phenylsulfonyl)ethylene adds to the same triene from below-plane has led more recently to an improved synthesis of [4]peristylane-2,4-dione (528) [280]. Epoxidation of 556 leads to 712 and sets the

stage for controlled reductive desulfonylation. Sensitized irradiation of the resulting diene epoxide 713 and exposure of the caged product (714) to periodic acid, as before, rapidly and efficiently delivers the hemispherical diketone.

Eaton and his coworkers have uncovered a useful way to break the symmetry of *cis*-bicyclo]3.3.0]octane-3,7-dione. The formation of 715 illustrates the technique [339].

Biscyclopentannulation of the same diketone provides the pair of tetraquinanediones 716 and 717. The first of these products has been carried on to [5]peristylane (718), the overall route being more convenient than that originally devised [339].

An intermediate in the 716 → 718 sequence happens to be diene dione 719. Treatment of this compound with the methyl Grignard reagent followed by chromium trioxide oxidation leads to transposed double enone 720 in which the double bonds are suitably arranged for photocyclization to the cyclobutane 721. The substance is easily obtained and appears (along with its desmethyl analogue) to be a choice precursor to the still unknown D_{3h}-$(4^3.5^6)$nonahedrane 722.

102

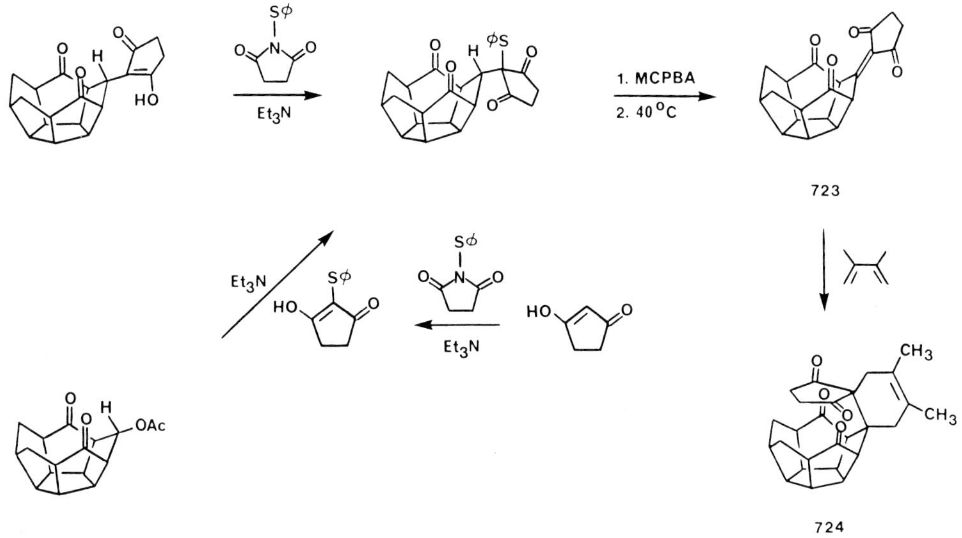

723

724

Methods for the synthesis of 2-alkylidene-1,3-cyclopentanediones have been developed and applied to the elaboration of 723, whose electron-deficient double bond reacts with 2,3-dimethylbutadiene to deliver 724 [340].

The readily available hexacyclic keto ether 725 undergoes thermal cracking as described earlier. The resulting dienone has been transformed into endo aldehyde 726, the crystalline dimethyl acetal of which is subject to intramolecular etherification upon exposure to phenylselenyl chloride. On reductive deselenation of this product, the novel trioxa[5]peristylane 727 emerges [341].

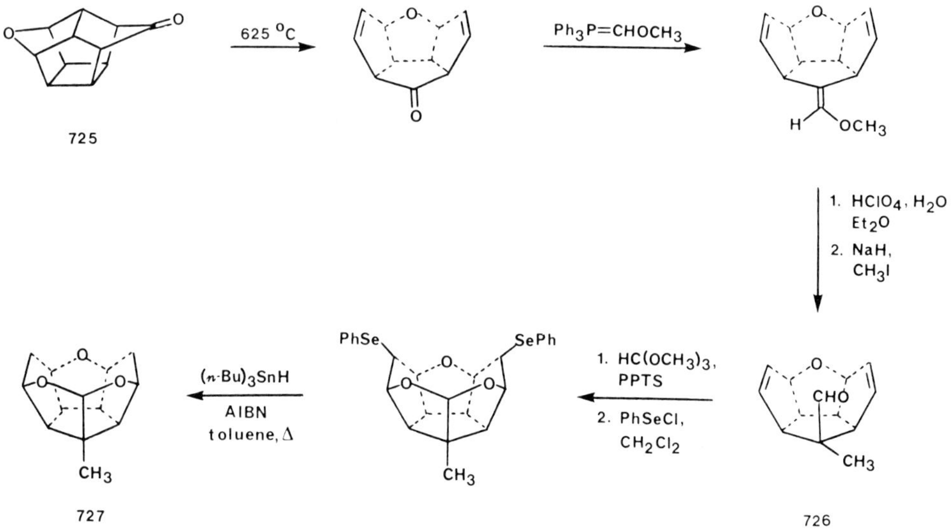

725

727

726

103

I Chemistry Surrounding Dodecahedrane

The development of alternative syntheses of pentagonal dodecahedrane continues to attract considerable attention. Among the more recent model studies is that based on the previously described C_{12}-oxa-tetraquinane 253. Treatment with a large excess of dichloroketene furnished 728, ring expansion of which with diazomethane and dechlorination afforded 729. Reaction of 729 with palladium dichloride gave a mixture of diene diones from which 730 could be isolated. On exposure to DBU, 730 smoothly epimerized to 731 whose hydrogenation yielded the spheroidal oxa-hexaquinanedione 732 [119, 342].

The Baldwin route to the C_{19}-heptaquinane 736 begins with the known tetracycle 85 (Scheme XLIV). Selective protection of the less hindered carbonyl group and cyclopentenone annulation of the two remaining reactive centers by the method of Trost led via 733 to 734. Hydrogenation of the double bonds in 735 could not be induced beyond the dihydro adduct, a molecule quite prone to internal Michael addition [343]. X-ray crystallographic analysis of several of the intermediates was utilized for structural assignment in many of these instances [344].

Prinzbach's pagodane approach takes advantage of the ease with which 737 undergoes [6 + 6] photocycloaddition to give 738 (Scheme XLV) [345]. This tetraene adds one equivalent of maleic anhydride stereospecifically. Hydrolysis of 739 to the diacid and the cuprous oxide-promoted decarboxylation provides diene 740 from which bis-(diazoketone) 741 is crafted. Photochemical Wolff rearrangement of 741 in aqueous tetrahydrofuran delivers diacid 742. Bis(iododecarboxylation) of this intermediate allows diiodide 743 to emerge and sets the stage for reduction to 744. Attempts are currently underway to effect the isomerization of pagodane to dodecahedrane (529).

Paquette has recently unveiled a preparatively useful dehydrogenative method for dodecahedrane synthesis [346]. Using an intimate 1:1 mixture of 5%

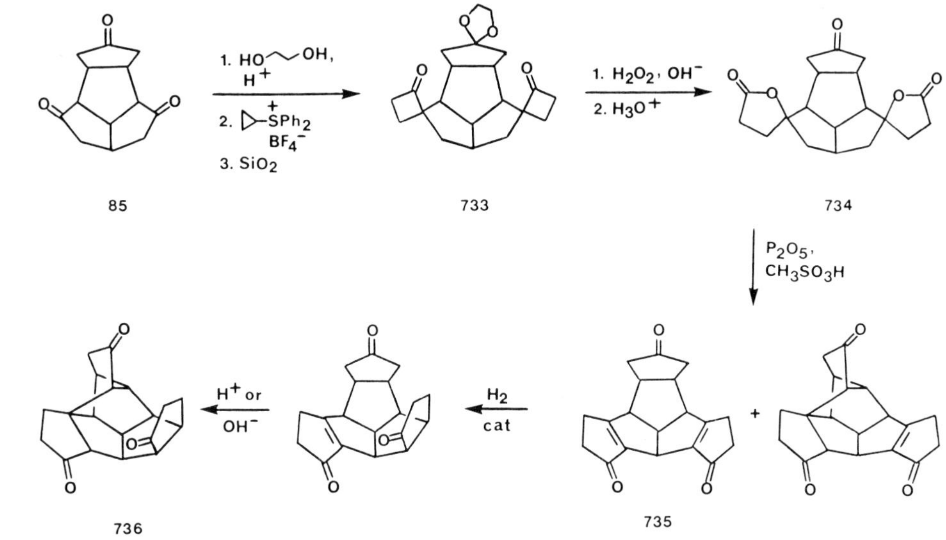

85 → **733** → **734**

736 ← **735**

Scheme XLIV

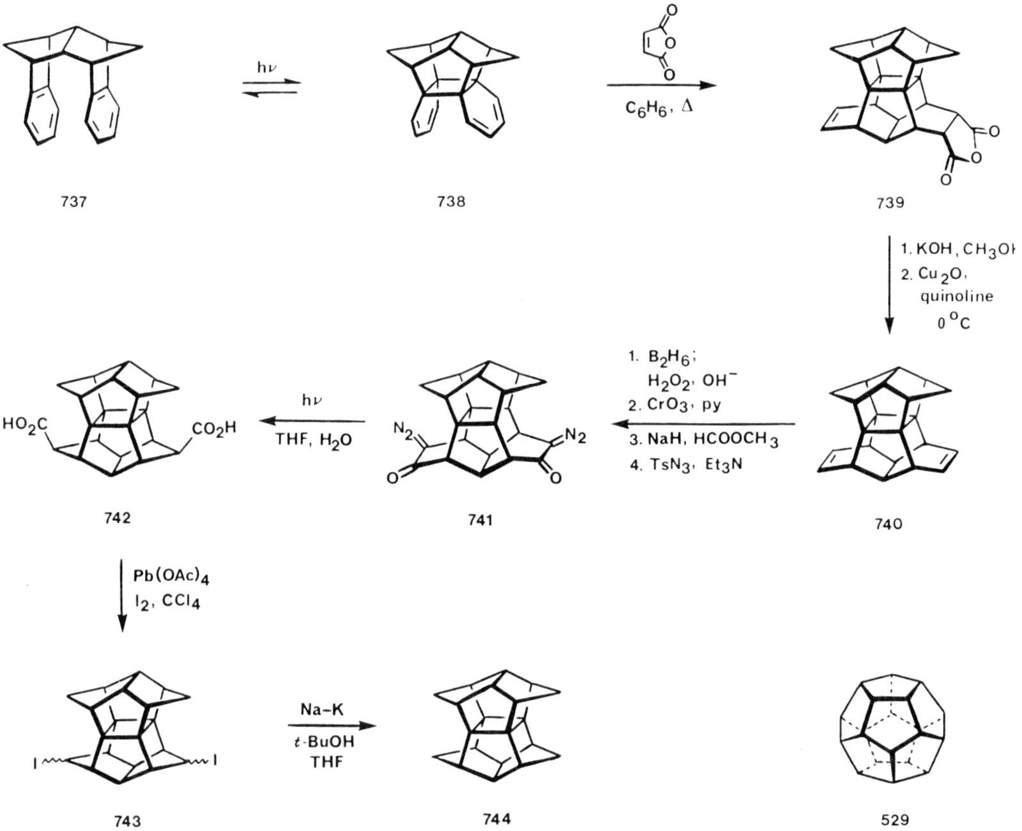

737 ⇌ **738** → **739**

742 ← **741** ← **740**

743 → **744**   **529**

Scheme XLV

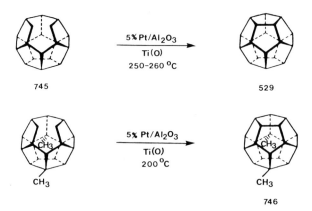

| 745 | 5% Pt/Al₂O₃ Ti(O) 250-260 °C | 529 |

Pt on alumina and finely divided Ti(O), the Ohio State group has found it possible to prepare the parent hydrocarbon in moderate amounts from seco derivative 745 and to obtain the previously unknown dimethyl derivative 746.

Although the mechanistic details of this process are lacking, it is already clear that demethylation can be made to occur and that a direct correlation exists with steric accessibility of the cleaved alkyl group to the catalyst. Thus, 747 has been transformed into 748 and 749 [346].

The catalyst system is also capable of carbocyclic 1,2-annulation. At 200 °C, 750 is transformed into benzyldodecahedrane (751); an incease in temperature to 260 °C triggers further dehydrocyclization to the indano derivative 752. Oxygen atoms need not interfere as seen by the fact that 754, which is available by Friedel-Crafts cyclization of 753, undergoes closure to 755 [346].

VI Natural Products Chemistry

A Isolation and Physical Properties

Tricyclic compounds 756 and 757 have been found to exist in low concentration in lavender oil. Structure proof was realized by selenium dioxide oxidation of α-cedrene. Hydrogenation of 756 and subsequent copper-catalyzed oxidation of the saturated aldehyde furnished 757 [347].

756   757

The diterpene hydroxyacid 758, characterized by X-ray crystallography, has been isolated from *Eremophila georgii* Diels [348]. Further characterization of 2-*epi*-jalaric acid, a lac resin constituent, has been made [349].

758   759

$\Delta^{9,12}$-Capnellene-2β,5α,8β,10α-tetrol (760) has been obtained from the alcyonarian *Capnella imbricata* and its three-dimensional features determined by X-ray methods [350]. Fresh colonies of this soft coral have more recently yielded

760

761a, $R_1 = R_2 = R_3 = R_5 = H$, $R_4 = OAc$
b, $R_1 = R_3 = R_5 = H$, $R_2 = OAc$, $R_4 = OH$
c, $R_1 = R_3 = R_5 = H$, $R_2 = R_4 = OAc$
d, $R_1 = R_2 = R_5 = H$, $R_3 = OAc$, $R_4 = OH$
e, $R_1 = R_2 = R_5 = H$, $R_3 = R_4 = OAc$
f, $R_1 = R_3 = H$, $R_2 = R_5 = OAc$, $R_4 = OH$
g, $R_1 = β\text{-}OAc$, $R_2 = R_5 = H$, $R_3 = OAc$, $R_4 = OH$
h, $R_1 = β\text{-}OAc$, $R_2 = R_5 = H$, $R_3 = R_4 = OAc$

the eight acetylated capnellenes 761 a—h [351]. They are believed to be the major sesquiterpenes present in the living animal.

The sesquiterpene antibiotic terrecyclic acid A (762) has been isolated from *Aspergillus terreus* [352—4] and its absolute stereochemistry established [355]. This diquinane can be recognized to be a ring-opened form of quadrone. Pentalenolactone F (763) is a new metabolite isolated from *Streptomyces* [356]. Pentalenolactones P (764) and O (765) have also been characterized in the last couple of years and join this fascinating class of natural products [357]. Subergorgic acid (766), a cardiotoxic sesquiterpene with an angular triquinane skeleton, has been readily purified from extracts of the western Pacific gorgonian *Subergorgia suberosa* [358]. An investigation of several strains of the basidiomycete *Crinipellis stipitaria* has led to the isolation of the two antibiotically active compounds crinipellin A (767) and crinipellin B (768) along with two inactive accompanying substances dihydrocrinipellin B (769) and tetrahydrocrinipellin A (770) [359].

762   763   764

765   766   767

768   769   770

Bohlmann's group continues to uncover polyquinane natural products at an unabated pace. These include the four isocomene derivatives 771 a—d and the dihydroxysilphinene isovalerate 772 from the roots of *Callilepsis salicifolia* [360], the presilphiperfolene diester 773 from the aerial parts of *Senecio anteuphorbium* [361], the nine isocedrene derivatives 774 a—c, 775 a—c, 776, and 777 a,b from the aerial parts of *Jungia stuebelii* [362], the keto silphinine 778 from *Dugaldia hoopesi* [363], and 13-isobutyryloxysilphinen-3-one (779) from *Dugaldia integrifolia* [364].

Deslongchamps has isolated the six new metabolites 780–784 from the extracts of *Ryania Speciosa*. These compounds are closely related to ryanodine, the known insecticidal toxic alkaloid from this plant [365].

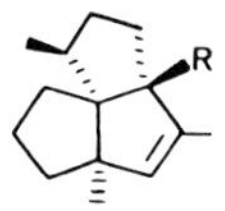

771a, R = CH₂OH
  b R = CH₂Oi-Val
  c, R = CHO
  d, R = COOCH₃

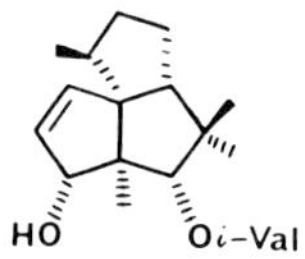

772

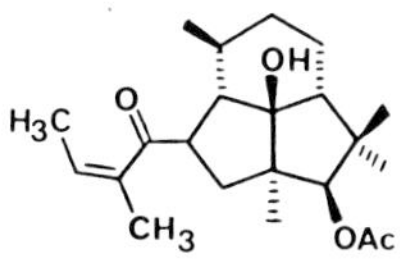

773

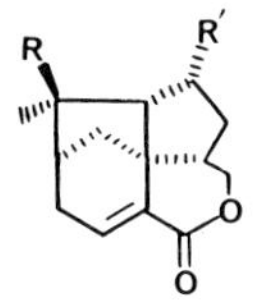

774a, R = COOH, R′ = H
  b, R = CH₃, R′ = OH
  c, R = CH₃, R′ = H

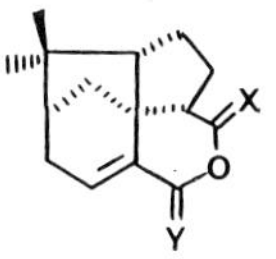

775a, X = β-OH, H; Y = O
  b, X = O; Y = H₂
  c, X = OH, H; Y = H₂

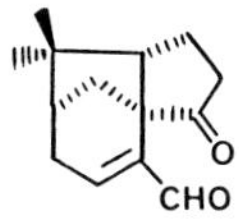

776

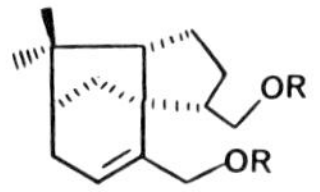

777a, R = Ac
  b, R = H

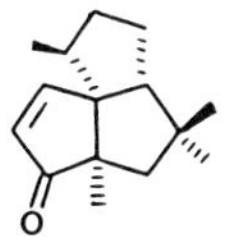

778

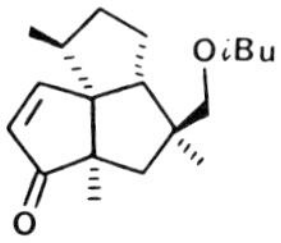

779

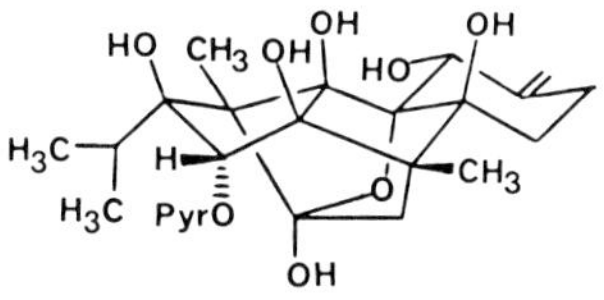

780

Pyr = pyranose

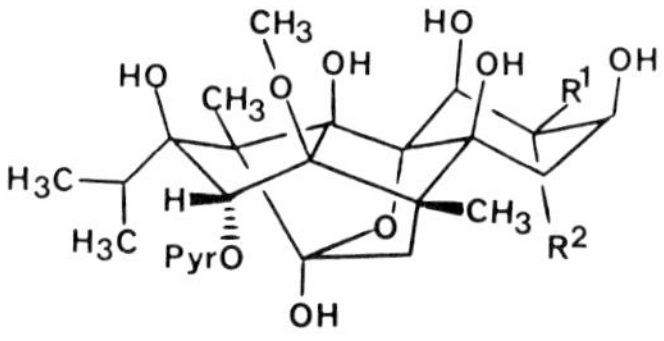

781a, R¹ = CH₃, R² = H
  b, R¹ = R² = CH₂

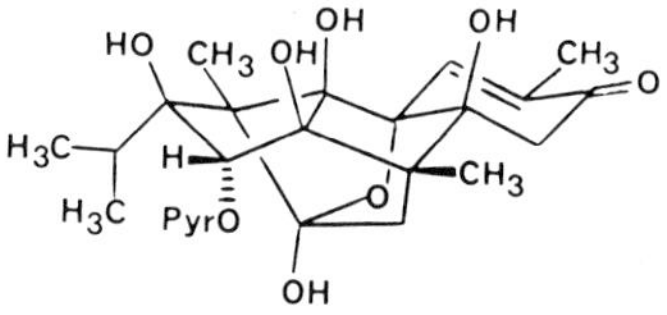

782

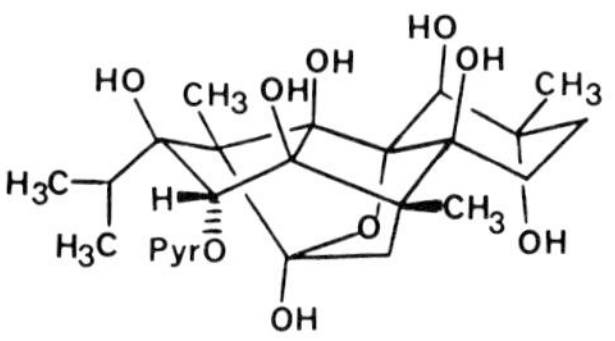

783

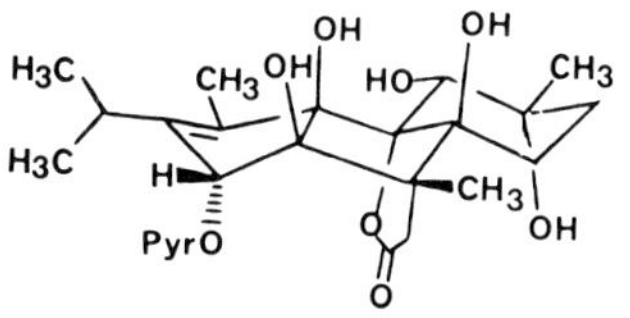

784

109

B Biosynthesis and Chemical Transformations

Cane and Tillman have found it possible to prepare a cell-free extract of *Streptomyces* that catalyzes the cyclization of *trans,trans*-farnesyl pyrophosphate to pentalenene [366]. Use of labelled substrate 785 gave pentalenene tritiated at its olefinic site, an event consistent with the illustrated sequence.

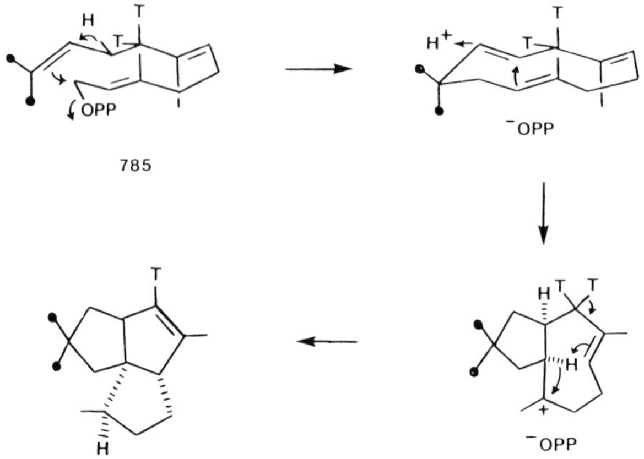

785

The biosynthesis of pentalenolactone (788) is presently considered to proceed from pentalenene via deoxypentalenic acid (786a), 786b, and 787 (Scheme LXVI) [367]. Pentalenolactone F (763) and most probably pentalenolactone E (788) are shunt pathway products.

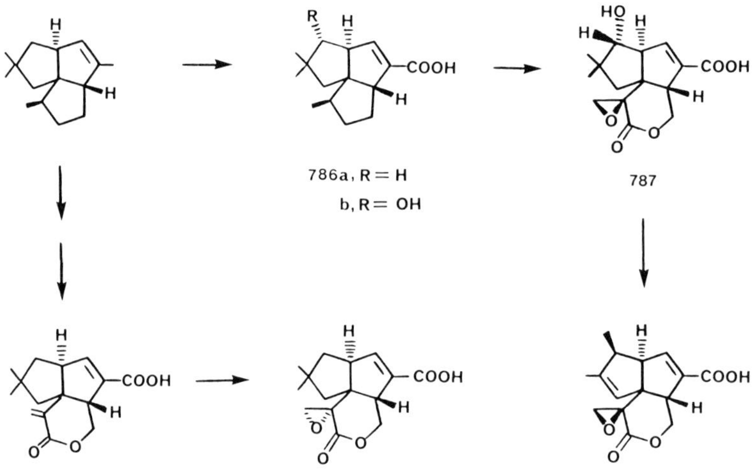

786a, R = H
b, R = OH

787

789   763   788

Scheme XLVI

Feedings of C-1 and C-2-labelled sodium acetate to *Aspergillus terreus* gave ^{13}C-enriched samples of quadrone and terrecyclic acid A (762). NMR analysis showed patterns of isotope incorporation that are consistent with formation of the

diquinane antibiotics by cyclization of farnesyl pyrophosphate [368]. Independent tritium and ^{13}C experiments in two other laboratories have led to the same conclusion [369, 370]. The specific antiviral activity of pentalenolactone O (765) has been detailed [371].

Laurenene (790), the solitary known example of a naturally occurring fenestrane, undergoes formic acid-induced rearrangement to 791. This structural assignment has been substantiated by X-ray analysis of the derived diol 792 [372].

790   791   792

Ozonolysis of 790 or autoxidation of laurenan-2-one (793) and subsequent methylation yields the keto ester 794 a, a substance sensitive to base-catalyzed epimerization to 794 b. Attempted enol acetylation, Meerwein-Ponndorf-Verley reduction, or high pressure hydrogenation of 794 a in the presence of Adams catalyst and either ferric chloride or perchloric acid caused diene 795 to emerge. This product forms most readily by treatment of 794 a with p-toluenesulfonic acid monohydrate in benzene at room temperature [373].

793   794 a, $R^1 = CH_3$, $R^2 = H$ 795
               b, $R^1 = H$, $R^2 = CH_3$

VII Synthesis of Nonpolycyclopentanoid Natural Products by Way of Diquinane Intermediates

Di- and triquinanes have increasingly found service as building blocks for construction of natural products lacking formal polycyclopentanoid frameworks [374]. In this segment of the review, the most recent applications of successful strategems of this type are surveyed.

A Allamcin

Allamcin (801) is a recently isolated member of the iridoid family of biologically active lactones found in *Allamanda* sp [375]. Taking advantage of the efficiency with which bicyclo[3.3.0]oct-7-en-2-one condenses with 2,4,6-triisopropylbenzenesulfonylhydrazine, Parkes and Pattenden treated this derivative with *n*-butyllithium in TMEDA and quenched the resulting vinyl anion with dimethylformamide to produce dienol 796 (Scheme XLVII) [376]. Following arrival at the enol acetate, treatment with buffered peracetic acid led to acetoxy aldehyde 797. The dianion of 2-phenylthiobutanoic acid acted on 797 to give four diastereomers of spirolactone 798 whose oxidation to vicinal diols permitted chromatographic

112

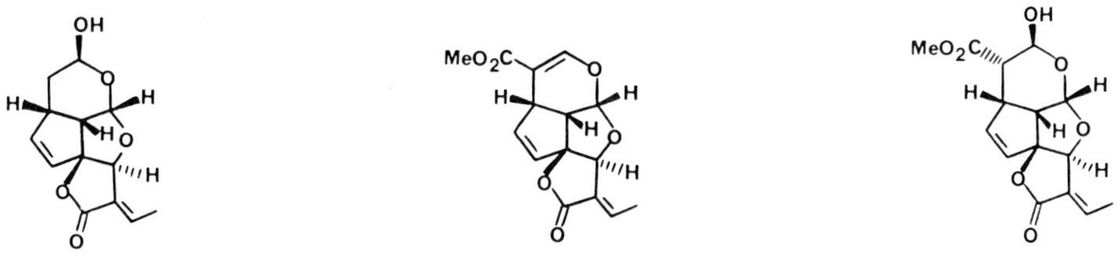

801   802   803

Scheme XLVII

separation of 799. Saponification followed by oxidation at sulfur and thermally induced elimination of the sulfoxide gave rise to triol 800. Oxidative cleavage of the 1,2-diol unit in 800 with concomitant cyclization furnished allamcin. Several alternative methods for proceeding from 799 to 801 proved serviceable. Since allamcin has previously been converted into plumericin (802) [377] and allamandin (803) [378], the present synthesis constitutes a formal preparation of these iridoids as well.

B Boonein

The monoterpene lactone boonein (809), isolated from the bark of the Nigerian tree *Alstonia boonei* DeWild [379], has been synthesized for the first time in seven steps from the bicyclooctenone 804 (Scheme XLVIII) [380]. Hydride reduction of 804 gave a mixture of alcohols dominated by 805. Protection of the hydroxyl group and hydroboration-oxidation furnished 806 and 807 in 16% and 69% isolated yields. Formation and ozonolysis of the silyl enol ether led uniquely to δ-lactone 808. Tin hydride reduction of this intermediate proceeded with inversion of configuration. Upon desilylation, (±)-boonein emerged.

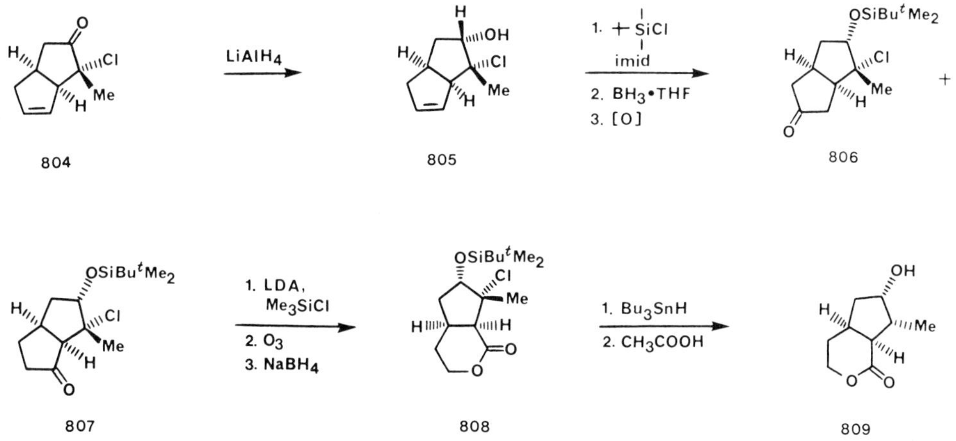

804   805   806

807   808   809

Scheme XLVIII

C Brefeldin A

A interesting synthesis of brefeldin A seco acid (815) through application of γ-hydroxyalkyl stannane oxidative fragmentation methodology has been developed by Nakatani and Isoe [381]. Their protocol (Scheme XLIX) involves initial attachment to 810 of the oxygenated pentyl and trimethylstannyl substituents as found in 811. Introduction of the (*E*)-vinyllithium reagent 812 set the stage for lead tetraacetate oxidation to give 813. Once adjustment of oxidation levels had been accomplished, base-catalyzed isomerization of 814 gave the targeted 815, an intermediate in an earlier total synthesis of brefeldin A.

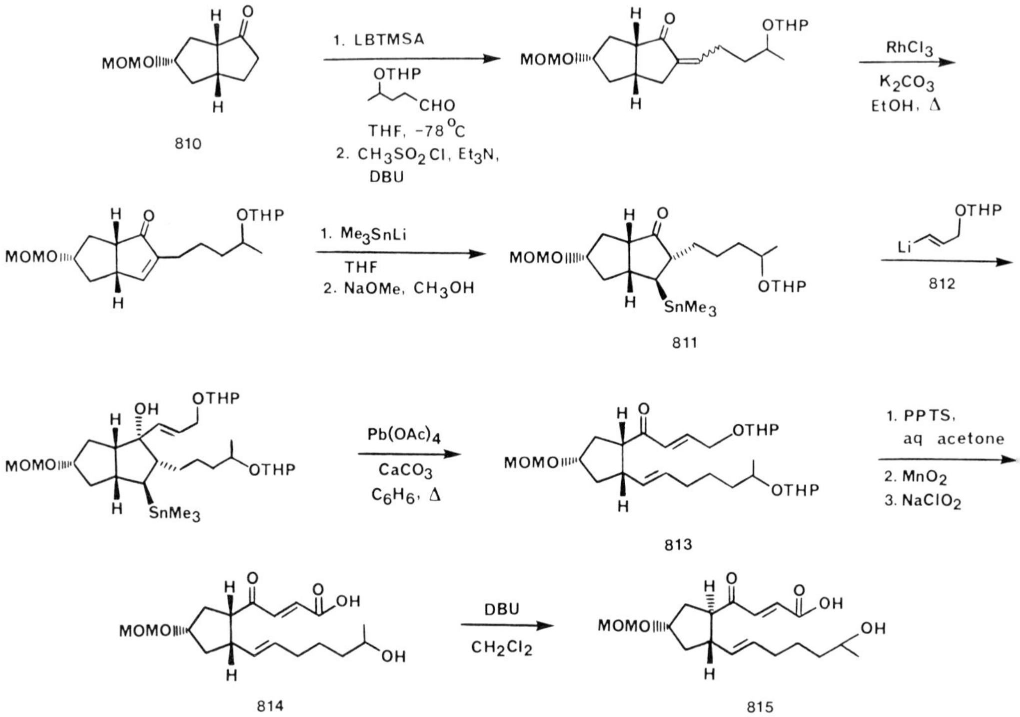

Scheme XLIX

D Dendrobine

A cationic rearrangement route involving an immonium intermediate has been developed as a potential path to the physiologically potent alkaloid dendrobine (822, Scheme L) [382]. Cyclopentenone 816 was prepared from *tert*-butyl acetoacetate in seven steps. Upon irradiation, this starting material underwent intramolecular [2 + 2] photocycloaddition to tricyclic ketone 817. The oximation product 818 was transformed in a six-step sequence into carbinol amide 819. Heating of formic acid solutions of this mixture at 100 °C permitted the isolation in 55% yield of 820 whose hydride reduction delivered amino alcohol 821.

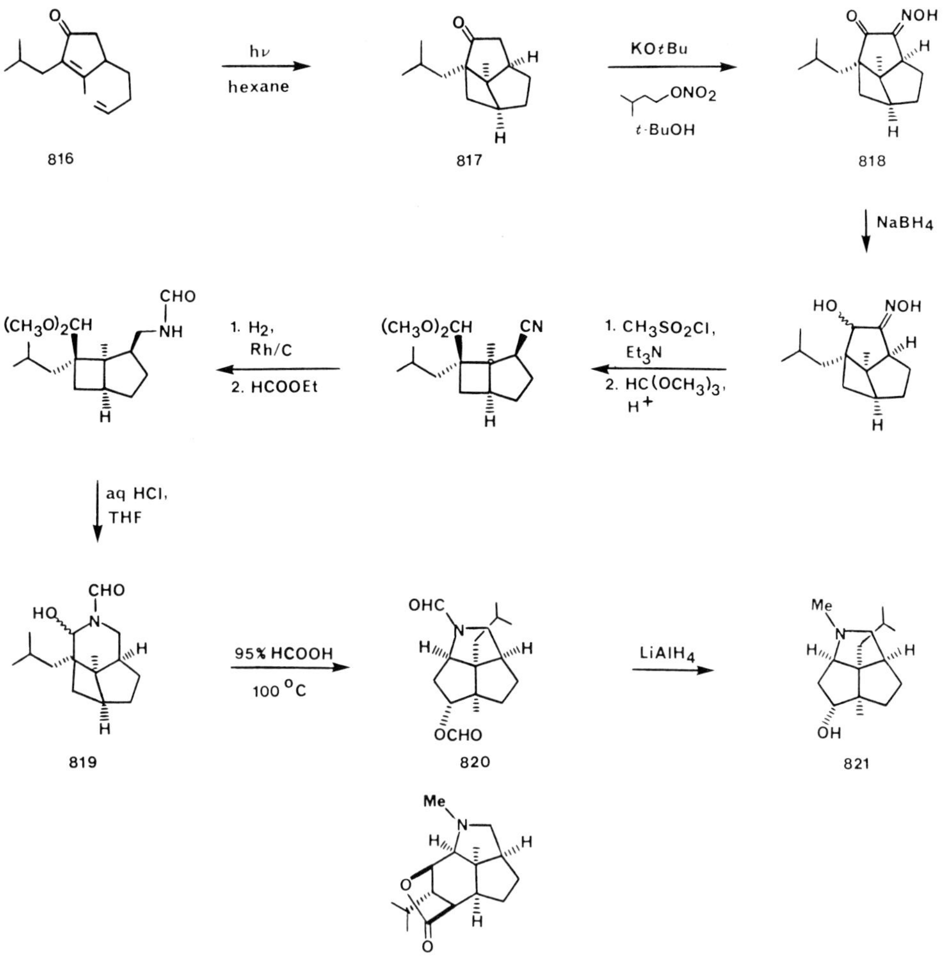

Scheme L

E Forsythide Aglycon Dimethyl Ester

Forsythide (827 a) is a naturally occurring iridoid glycoside that has been isolated from fresh leaves of *Forsythia irridissima* Lindl [383]. To arrive at its aglycone dimethyl ester (827 b), Takemoto and Isoe effected the methoxycarbonylation of tricyclo[3.3.0.0^{2,8}]octan-3-one in order to guarantee exclusive cleavage of its C2—C8 bond during subsequent acid-catalyzed cyclopropane ring cleavage (Scheme LI) [384]. Subsequent hydrolysis afforded hydroxy keto ester 823. Removal of the ketone carbonyl oxygen set the stage for Jones oxidation and the isolation of 824. The derived cyanohydrin was dehydrated to give unsaturated nitrile 825. This intermediate was hydrolyzed, treated with diazomethane, and deconjugated in conventional fashion. Ozonolysis of 826 followed by reductive work-up led to 827 b in stereocontrolled fashion.

115

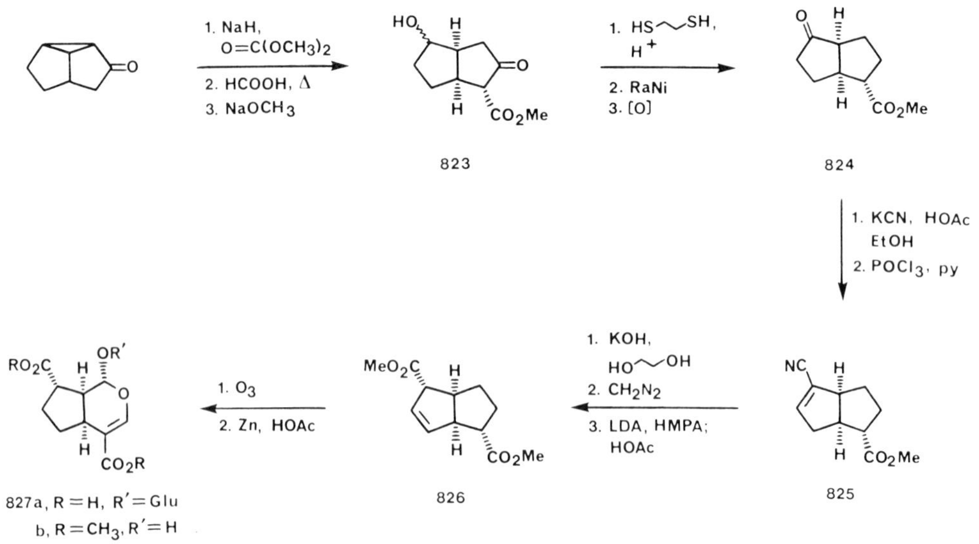

823

824

827a, R=H, R'=Glu
b, R=CH₃, R'=H

826

825

Scheme LI

F Ikarugamycin

The Boeckman approach to ikarugamycin (832) entails Wittig condensation of phosphorane 828 with aldehyde 829; following exposure of the unpurified product to iodine, the pure all-trans triene 830 was isolated [385]. This substance underwent smooth Diels-Alder cyclization at 140 °C to provide 831 as the major stereoisomer. Following Dibal reduction, closure of the final bond was effected by acidic ketal hydrolysis, conversion to the tosylate, and base-promoted cyclization.

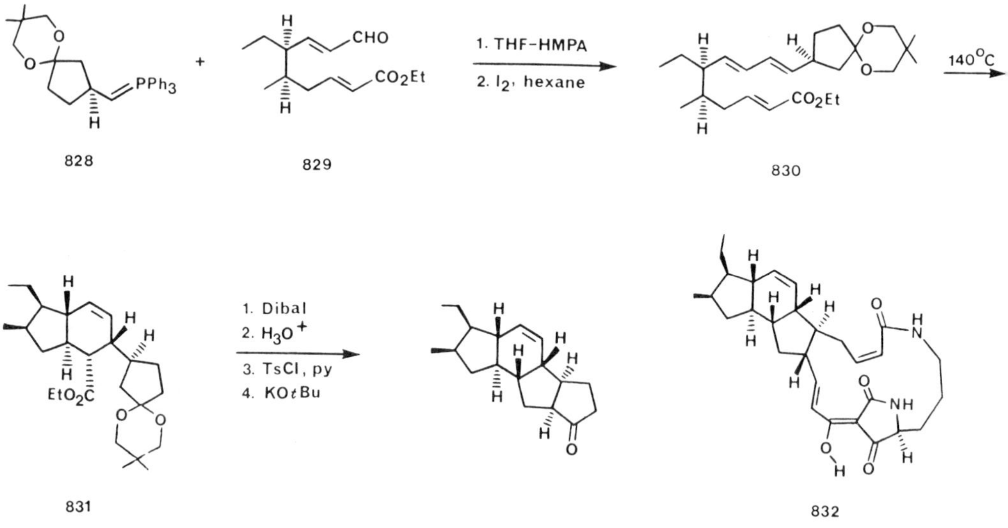

828

829

830

831

832

116

G Iridodial

This monoterpene (836) has been prepared in its dextrorotatory form beginning with regioselective methylation of (—)-tricyclo[3.3.0.0^{2,8}]octan-3-one [386]. Reduction of 833 with Dibal afforded predominantly (91%) the endo alcohol, which was transformed stereoselectively (83%) into 834 upon reaction with methanesulfonyl chloride and triethylamine. Treatment with iodide ion transformed 834 into the β-iodide. This intermediate reacted with methylmagnesium iodide and cuprous iodide to effect replacement by methyl with retention of configuration. Hydrocarbon 835 was bishydroxylated and finally subjected to periodate cleavage to give (+)-836.

833 → 834 → 835

(+)-836

H Isoiridomyrmecin

The arene-olefin cycloaddition reaction has provided a straightforward approach to isoiridomyrmecin (841) [387]. Photolysis of benzene and vinyl acetate produced 837 in low yield as the main isolable product. Subsequent to reductive cleavage of the acetate functionality and oxidation, methylation under kinetic conditions was implemented to produce 838 with complete selectivity. Dimethylcuprate addition to 838 and direct trapping of the enolate gave rise chiefly to 839. Completion of the synthesis required selective hydrogenation and subsequent

837 → 838 → 839

840 → 841

117

ozonolysis in methanol with a sodium borohydride work-up. This led to 840 whose reduction with sodium cyanoborohydride in acidified aqueous tetrahydrofuran provided 841.

I Loganin and Analogues

The iridoid loganin (845) is a key biosynthetic intermediate for indole, monoterpene alkaloids, and other natural products. This relatively simple substance has accordingly been the target of a considerable amount of synthetic activity. For example, Hewson and MacPherson effected the decarbomethoxylation of 27 a to give 842 and ultimately keto ester 843 (Scheme LII) [21]. Following stereospecific borohydride reduction, the hydroxyl stereochemistry was inverted and the unsaturated ester functionality deconjugated to give 844. This substance had previously been transformed into loganin [388].

In the same study, 27 a was treated with methyllithium to deliver alcohol 846 stereospecifically. Hydrolysis gave keto alcohol 847, a compound which had earlier been converted to chrysomelidial (848) [389].

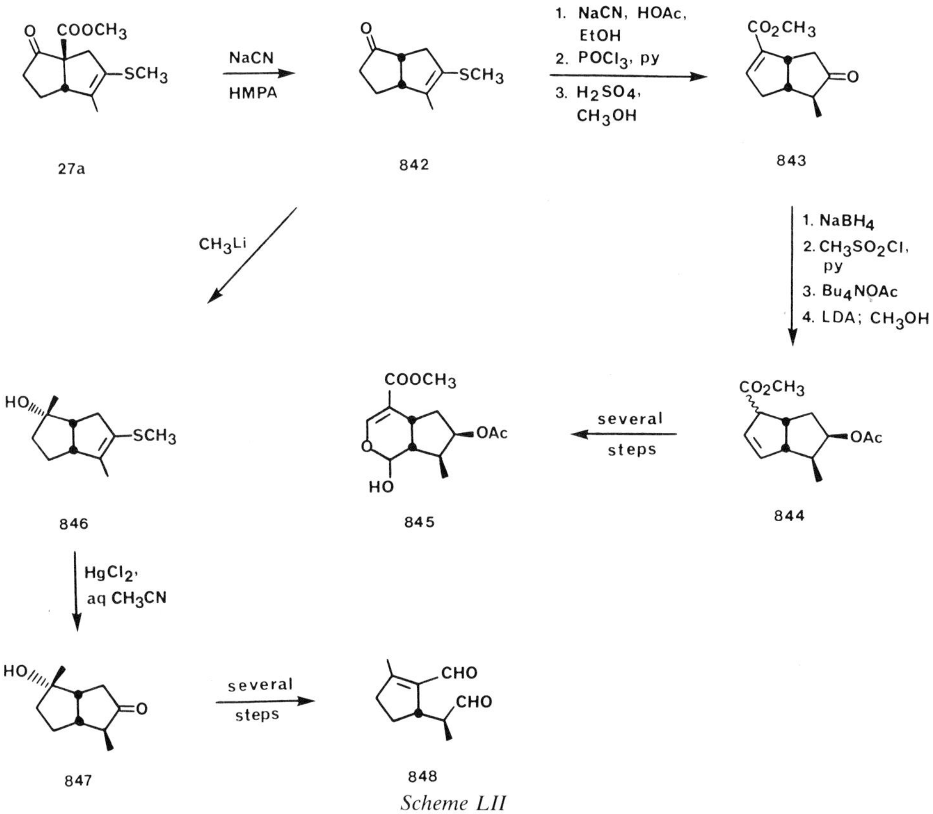

Scheme LII

Kon and Isoe chose to begin their synthesis with tricyclooctanone 833 and to effect conversion to 849 through ring cleavage, methanolysis, and acetalization

[390]. This major isomer was next oxidized and subjected to the Shapiro reaction. Trapping of the vinyl anion with methyl chloroformate gave α,β-unsaturated ester 850. Once again, deconjugation proved to be necessary. Following osmylation and deketalization, dihydroxy ketone 851 emerged. Oxidation of this intermediate with sodium periodate afforded 7,0-didehydrologanin aglycone, which was converted into the 0-methyl derivative 852, a synthetic precursor to loganin [391].

Guilard's stereoselective synthesis of loganin precursor 855 centered around carbomethoxylation of 853 and subsequent Shapiro degradation to give β,γ-unsaturated ester 854 directly [392]. Deketalization and ozonolytic cleavage of this intermediate led to 855.

Levorotatory tricyclic ketone 833 has served as the starting point for the preparation of $(+)$-859, the 6-acetate of loganin aglucone [393]. Isomerization of 833 with Nafion-Me$_3$Si (a perfluorated trimethylsilyl sulfonate resin) and subsequent borohydride reduction delivered 856 exclusively. Its inverted acetate underwent TiCl$_4$-catalyzed ene reaction with diethyl oxomalonate to give 857. Degradation to ester 858 set the stage for conversion to 859 by conventional methodology.

Palladium-catalyzed cycloaddition of 860 to 2-cyclopentenone also creates the proper substitution pattern of loganin. Application of the Shapiro process to 861 followed by carboxylation led to 862. Deconjugation of the enoate and double bond cleavage gave rise to the keto form of loganin aglucone (863) [394].

The Vandewalle synthetic effort has given rise to the aglucones of 8-epiloganin (870) and mussaenoside (869) [395]. The availability of 864 nicely set the stage for a nitrone-olefin cycloaddition reaction and formation of 865 (Scheme LIII). With compound 866 in hand, it proved an easy matter to prepare keto ester 867. Mercuric acetate-mediated addition of water to the derived diene ester 868 enabled subsequent ready access to 869. On the other hand, hydroboration of 868 made possible the acquisition of 870.

J Oplopanone

Piers and Gavai noted that addition of the cuprate derived from 871 to several 2-cyclopentenones and to 872 resulted in formation of the (Z)-ethylidenecyclopentane annulation products 873 and 874, respectively [396]. This methodology was then extended to preparation of oplopanone (875) and several related sesquiterpenoids.

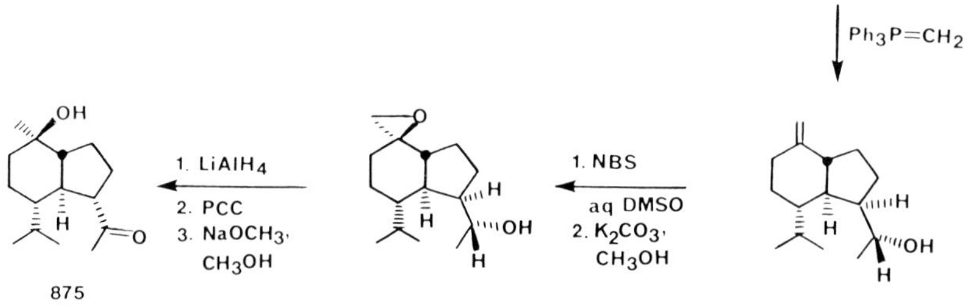

875

K Plumericin

Plumericin, an antifungal, antibacterial, and antitumor substance of some note, has been shown to possess the densely functionalized structure 882. A biomimetic

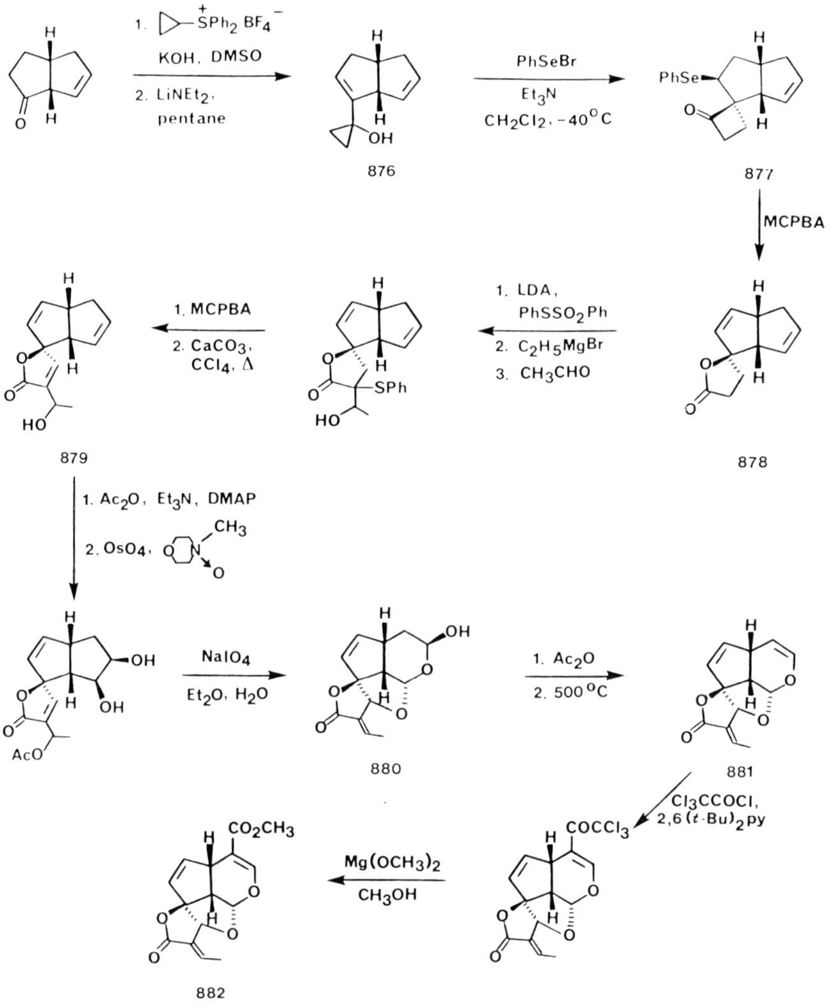

876

877

879

878

880

881

882

Scheme LIV

approach to 882 due to Trost is founded on an operating strategy involving conjugate addition-elimination [377]. Following arrival at 876, the stereochemistry of the target is set into place by phenylselenyl bromide-initiated ring expansion. Double oxidation of 877 provides diene lactone 878 directly (Scheme LIV). For elaboration of butenolide 879, the magnesium enolate of the sulfenylated lactone was generated and condensed with acetaldehyde. The hydroxyl group in 879 was primed for its role as a leaving group by acetylation. Regiospecific cis hydroxylation and subsequent periodate cleavage sufficed to promote cyclization of the appropriate diastereomer to 880. Pyrolysis of the acetate of 880 delivered dihydropyran 881, the requisite carbomethoxylation of which was achieved in clever fashion by trichloroacetylation and ensuing methanolysis.

L Trichodiene

Trichodiene (888), the parent hydrocarbon of the tricothecane class of sesquiterpenoids, has recently been prepared by means of a convergent strategy involving combination of a simple cyclopentane derivative with a substituted cyclohexene [397]. Thus, Friedel-Crafts acylation of 1,4-dimethylcyclohexene with 883 and subsequent Nazarov cyclization gave rise to tricyclic ketone 884 as a 2.4:1 mixture of the cis-anti-cis and cis-anti-trans diastereomers. Since the stereocenters other than at the quaternary carbons are immaterial to the synthesis, the mixture was directly oximated and cleaved by second-order Beckmann fragmentation. The resulting nitriles 885 and 886 were reduced sequentially with lithium aluminum hydride and with lithium in liquid ammonia to give amine 887. Pyrolysis of the amine oxide provided homogeneous racemic 888.

M Udoteatrial

The antimicrobial diterpene udoteatrial, isolated from the alga *Udotea flabellum*, was first isolated and characterized in 1981 [398]. The assignment of structure as 889 was based on spectral analysis of its diacetate 890, since 889 expectedly interconverts readily with its stereoisomers. In the Whitesell synthesis of 889,

889   890

diene 891 was regio- and stereospecifically hydroborated with disiamylborane to give endo alcohol 892 (Scheme LV) [399]. An oxidation-(phenylseleno)lactonization sequence provided lactone 893, which was reduced and subjected to Wittig olefination. Claisen rearrangement of acetate 894 b as the *tert*-butyldimethylsilyl ketene acetal afforded silyl ester 895, a convenient progenitor of 896. Cleavage of both π-bonds in 896 with ozone in methanol released the three aldehyde groups, cyclization of which occurred spontaneously to form the cyclic mixed acetal

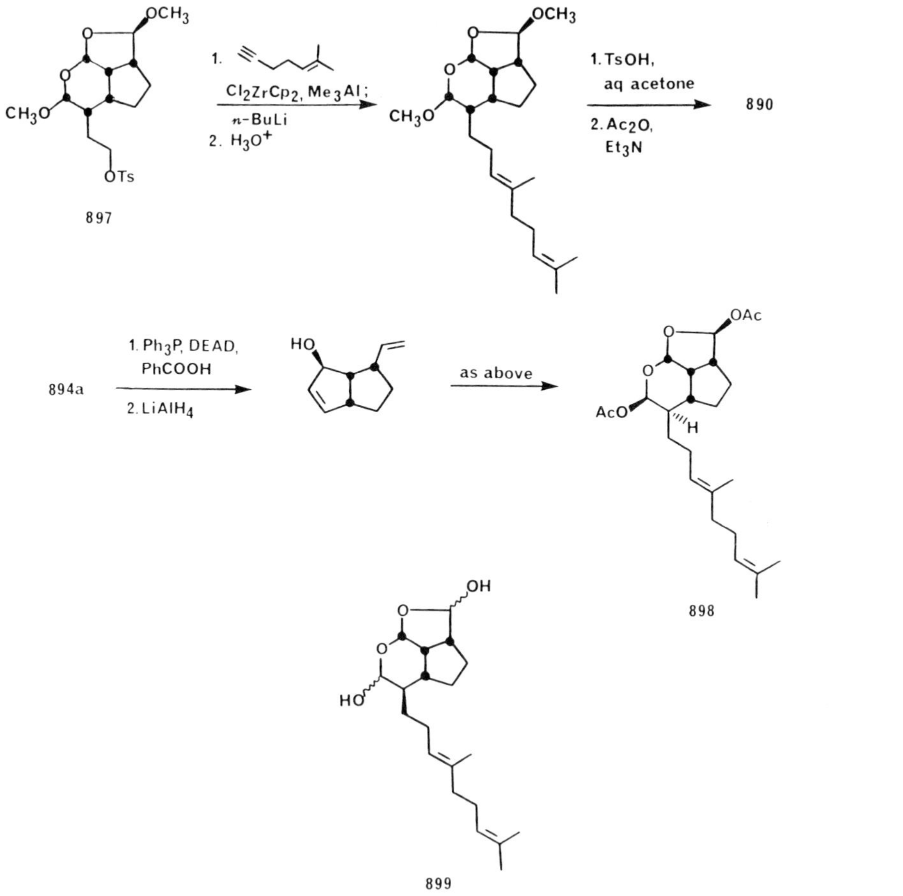

Scheme LV

system present in 897. Attachment of the dienyl sidechain was achieved by displacement of the tosylate with a zirconium-based ate complex. Hydrolysis of the acetals followed by acetylation afforded a major diacetate which was not identical spectroscopically with the substance derived from the natural product.

Since the present synthesis of 890 was unambiguous, the original structural assignment was considered erroneous. Accordingly, the C—7 epimer 898 was synthesized by returning to 894 a, inverting its hydroxyl stereochemistry by a Mitsunobu reaction and hydride reduction. The sequence of steps just described was repeated, affording ultimately 898. Consequently, udoteatrial has the structure depicted by 899.

N Xylomollin

An attempt to prepare the iridoid sesquiterpene xylomollin (907) has been pursued starting from keto ketal 900 [400]. Its Baeyer-Villiger oxidation led to 901 from which dialdehyde 902 was crafted (Scheme LVI). Treatment of this

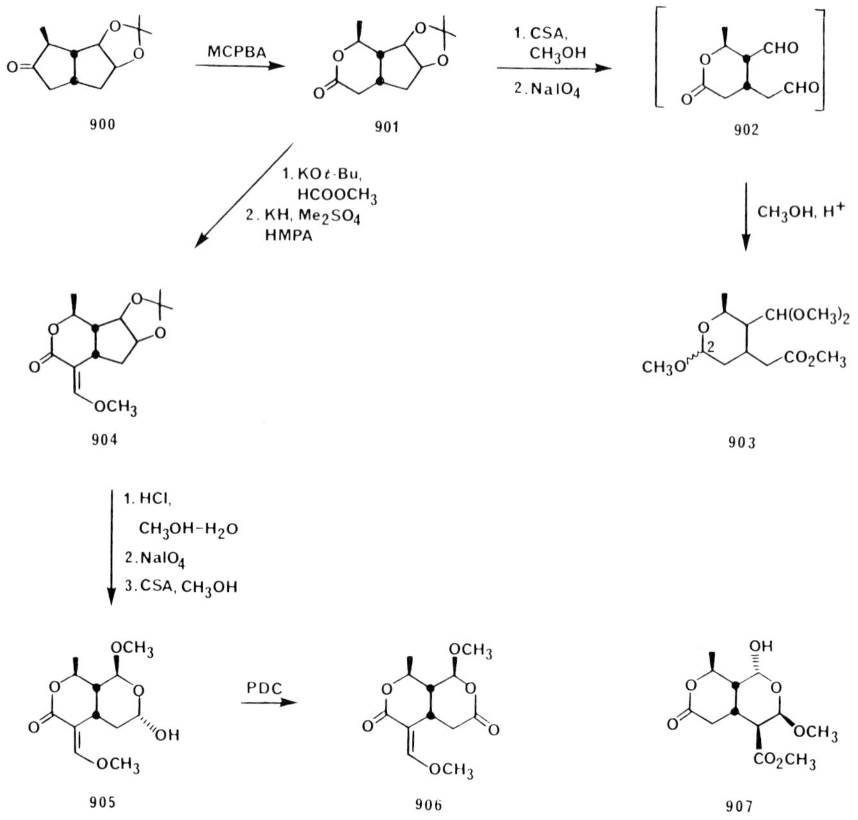

Scheme LVI

intermediate with acidic methanol led to lactolide ester 903 as an anomeric mixture at C-2. To arrive at xylomollin, the need exists to incorporate an additional framework carbon and the decision was made to accomplish this directly on 901. Consequently, methoxymethylidenation was effected, the glycol unit in 904 was selectively deprotected, and periodate cleavage furnished 905. Lactone 906, subsequently obtained by oxidation, contains all of the stereochemical relationships and basic functionalities present in xylomollin. However, it has not yet proven possible to effect the necessary interconnection.

VIII Synthesis of Diquinane Natural Products

A Cedranoids

In recent years, the tricyclic cedranoid sesquiterpenes have continued to be the subject of numerous synthetic studies. The novel carbocyclic framework of this group of natural products has resulted in a wide variety of synthetic approaches.

A formal synthesis of ($\pm$)-cedrene (910) developed by Rao [401] centered on the Dieckmann cyclization of the cyclopentane derivative 908. Subsequent

decarboxylative hydrolysis and reduction afforded ($\pm$)-norcedrenedicarboxylic acid (909). The conversion of 909 to ($\pm$)-cedrene has previously been reported by Stork [402]. Pattenden and Horton have reported an interesting synthesis of ($\pm$)-cedrene featuring sequential inter- and intramolecular Michael reactions [403, 404]. Thus, intermolecular Michael addition of the enolate derived from the key bicyclo[3.3.0]octanone intermediate 911 to 2-nitrobut-2-ene led to a diastereoisomeric mixture of the nitroketones 912 (Scheme LVII). Hydrolysis using sodium nitrite and n-propyl nitrite in dimethylsulfoxide afforded the 1,4-dione 913, which upon treatment with potassium $tert$-butoxide in $tert$-butanol underwent an intramolecular Michael reaction to give the tricyclo[5.3.1.0^{1,5}]undecanedione 914. Standard transformations led to the axial carbinol 915 which was subjected to Simmons-Smith conditions followed by hydrogenation and dehydration to afford ($\pm$)-cedrene (910) and ($\pm$)-2-epi-methyl-α-cedrene (916).

The synthesis of ($\pm$)-8(S),14-cedranediol (921) due to Landry (Scheme LVIII) features the intramolecular Diels-Alder reaction of an alkenylcyclopentadiene prepared in $situ$ by the retro Diels-Alder reaction of the corresponding alkenyldicyclopentadiene [405]. Thus, a number of standard transformations

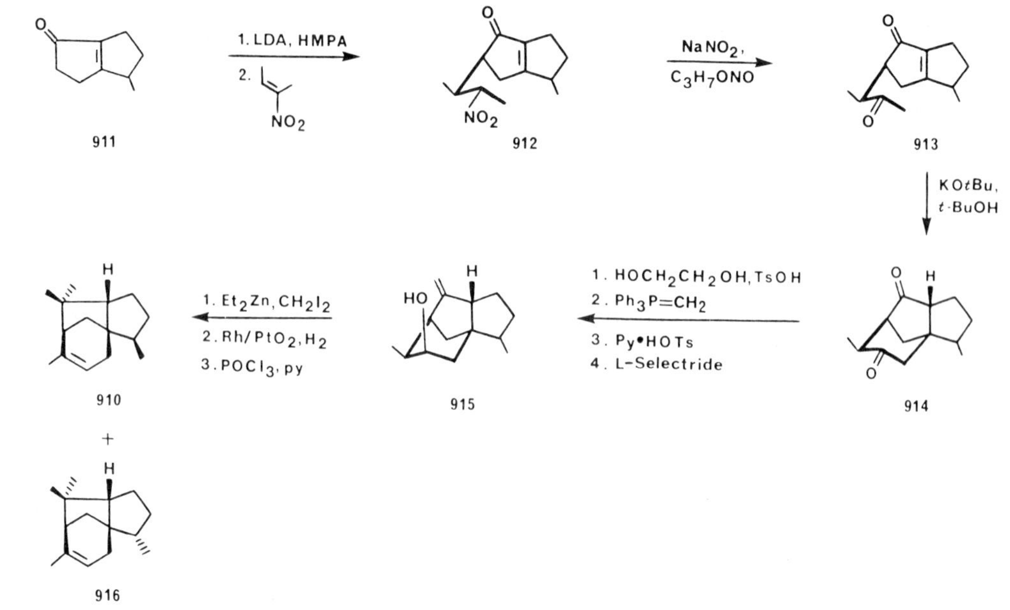
1. LDA, HMPA
2.
NO2
911
912
NaNO2,
C3H7ONO
913
KOtBu,
t-BuOH
914
1. HOCH2CH2OH, TsOH
2. Ph3P=CH2
3. Py•HOTs
4. L-Selectride
915
1. Et2Zn, CH2I2
2. Rh/PtO2, H2
3. POCl3, py
910
+
916

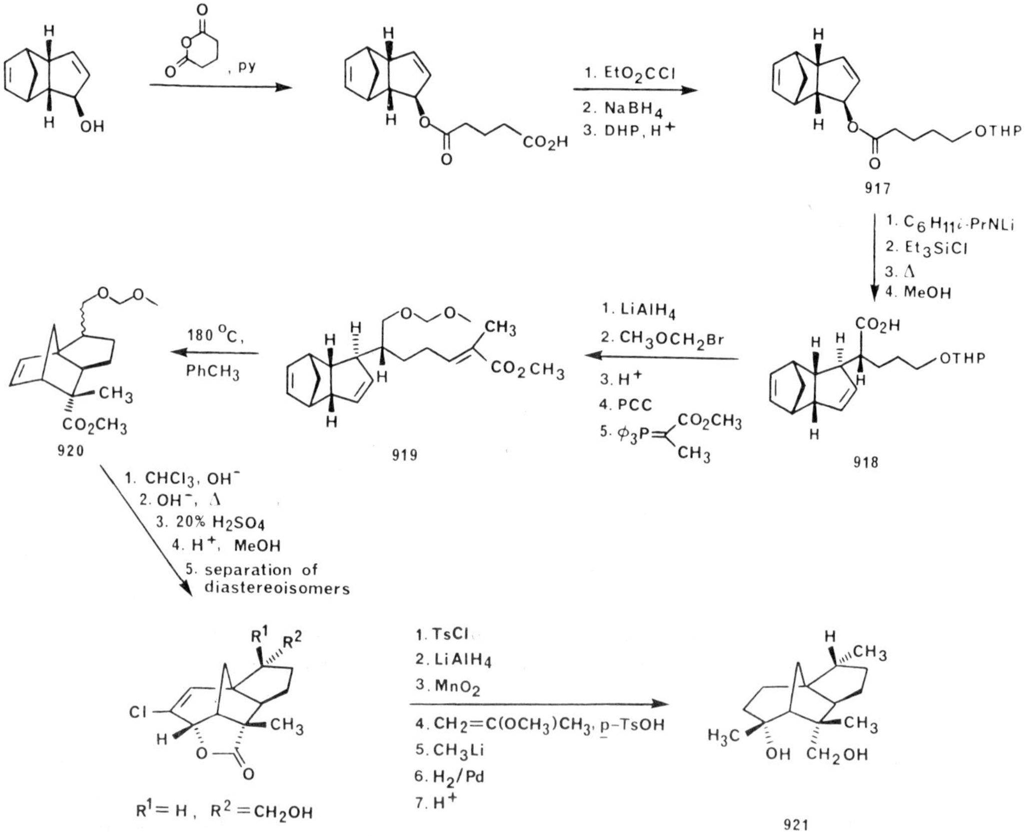
py
1. EtO2CCl
2. NaBH4
3. DHP, H+
CO2H
OTHP
917
1. C6H11i-PrNLi
2. Et3SiCl
3. Δ
4. MeOH
918
1. LiAlH4
2. CH3OCH2Br
3. H+
4. PCC
5. φ3P=C(CO2CH3)CH3
919
180 °C,
PhCH3
920
1. CHCl3, OH⁻
2. OH⁻, Δ
3. 20% H2SO4
4. H+, MeOH
5. separation of diastereoisomers
R1 = H, R2 = CH2OH
1. TsCl
2. LiAlH4
3. MnO2
4. CH2=C(OCH3)CH3, p-TsOH
5. CH3Li
6. H2/Pd
7. H+
921

provided access to intermediate 917 from dicyclopentadiene [406]. On heating, the resulting allyl silyl vinyl ether underwent the Claisen rearrangement which after hydrolysis afforded carboxylic acid 918. The key compound 919 was heated at 180 °C in toluene to effect a novel one-pot retro Diels-Alder reaction, [1,5]-sigmatropic rearrangement and intramolecular Diels-Alder reaction, and produced a diastereomeric mixture of esters 920. Dichlorocarbene ring expansion followed by basic hydrolysis and lactonization afforded the tricyclic carbon framework of the natural product.

In a formal synthesis of (±)-cedrene, Irie and his coworkers chose to employ an intramolecular Wadsworth-Emmons annulation reaction to construct the bicyclo[3.3.0]octenone framework [407]. To this end, the cyclopentanone triester 923 b obtained in low yield from the tri-*tert*-butylester 922 was converted to the corresponding phosphonate. Subsequent treatment with sodium hydride effected smooth cyclization to the bicyclo[3.3.0]octenone 924. A series of standard reactions led to (±)-norcedrenedicarboxylic acid (909), which has earlier been converted to (±)-cedrene by Stork (Scheme LIX) [402].

Recently Delmond and Adrianome have developed a general route to allyl-stannane [408, 409] and allylsilane [409] derivatives of various terpenic compounds and illustrated their isomerization under aqueous acidic conditions. Thus, α-cedrene has been converted to derivatives 925 and 926. The first compound upon treatment with acid gave the less stable exocyclic methylene isomer 927.

926

925

925   HCl / MeOH − 4% H₂O   927

B The Pentalenolactone Group

The *Streptomyces* antibiotic pentalenolactone (788), first isolated in 1970 [410], was demonstrated to be biosynthetically derived from humulene via several intermediates closely related to 788 [411]. The interesting biological properties of pentalenolactone have aroused considerable interest in its synthesis and that of a number of its biosynthetic precursors.

In this connection, Matsumoto and his coworkers have devised novel syntheses of pentalenolactones E, F, G, H and parent compound 788 by biogenetic-like cyclization of humulene. Thus bicyclic ketone 928a, derived from humulene [412], underwent transannular cyclization to the key triquinane 929 upon treatment with formic acid at 45 °C followed by sodium carbonate in methanol-water solution. After functional group manipulation, the α-methylene lactone was constructed via oxidation of the silyl dienyl ether 930. Pentalenolactone E methyl ester (931) was then directly accessible after sodium borohydride reduction and lactonization. Pentalenolactone E (933) and the F methyl ester (932) were easily obtained from 931 (Scheme LX) [413]. Pentalenolactone itself (788) and the H and G methyl esters (936 and 937) were synthesized via a similar route starting from the benzoate compound 934 (Scheme LXI) [414]. Cane and Thomas in their synthesis of the methyl esters of pentalenolactones E and F (931 and 932) effected the closure of the fused δ-lactone ring via an intramolecular insertion of an α-acyl

928a, R = ⸌H /⸌Me   83%

b, R = ⸌Me /⸌H   7%

929

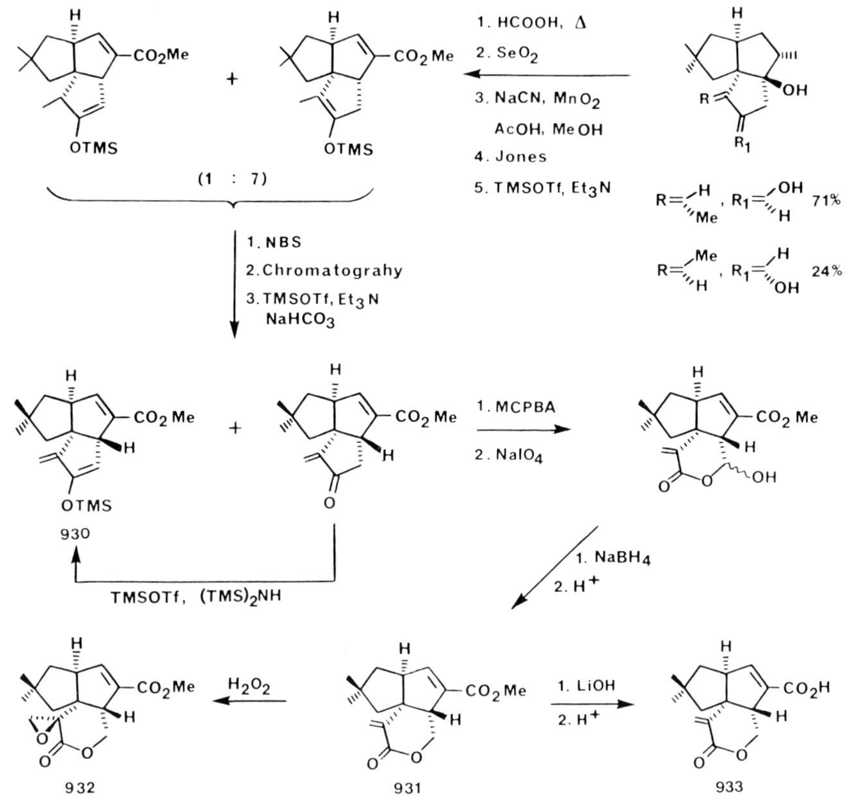

The reaction scheme shows various structures and reagents:

Top row: two bicyclic structures with CO₂Me and OTMS groups (1 : 7), plus a structure with OH, R, R₁ groups.

Reagents:
1. HCOOH, Δ
2. SeO₂
3. NaCN, MnO₂ AcOH, MeOH
4. Jones
5. TMSOTf, Et₃N

$R = \overset{H}{\underset{Me}{}}$, $R_1 = \overset{OH}{\underset{H}{}}$ 71%

$R = \overset{Me}{\underset{H}{}}$, $R_1 = \overset{H}{\underset{OH}{}}$ 24%

1. NBS
2. Chromatograhy
3. TMSOTf, Et₃N NaHCO₃

930 (OTMS structure) + (ketone structure)

1. MCPBA
2. NaIO₄

(structure with CO₂Me, lactone, OH)

TMSOTf, (TMS)₂NH

1. NaBH₄
2. H⁺

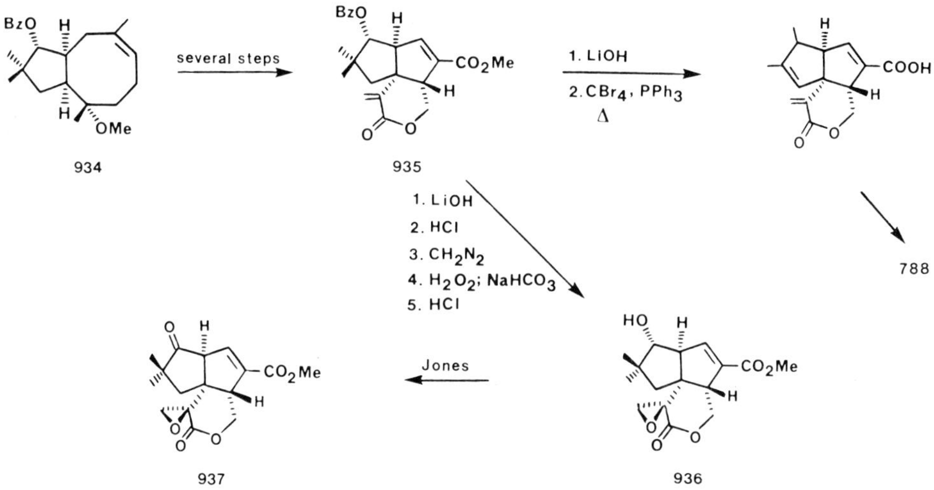

932 ← H₂O₂ ← **931** → 1. LiOH 2. H⁺ → **933**

Lower scheme:

934 (BzO, OMe structure) → several steps → **935** (BzO, CO₂Me lactone)

1. LiOH
2. CBr₄, PPh₃ Δ

→ **788** (COOH structure)

From 935:
1. LiOH
2. HCl
3. CH₂N₂
4. H₂O₂; NaHCO₃
5. HCl

→ **936** (HO, CO₂Me structure)

937 ← Jones ← **936**

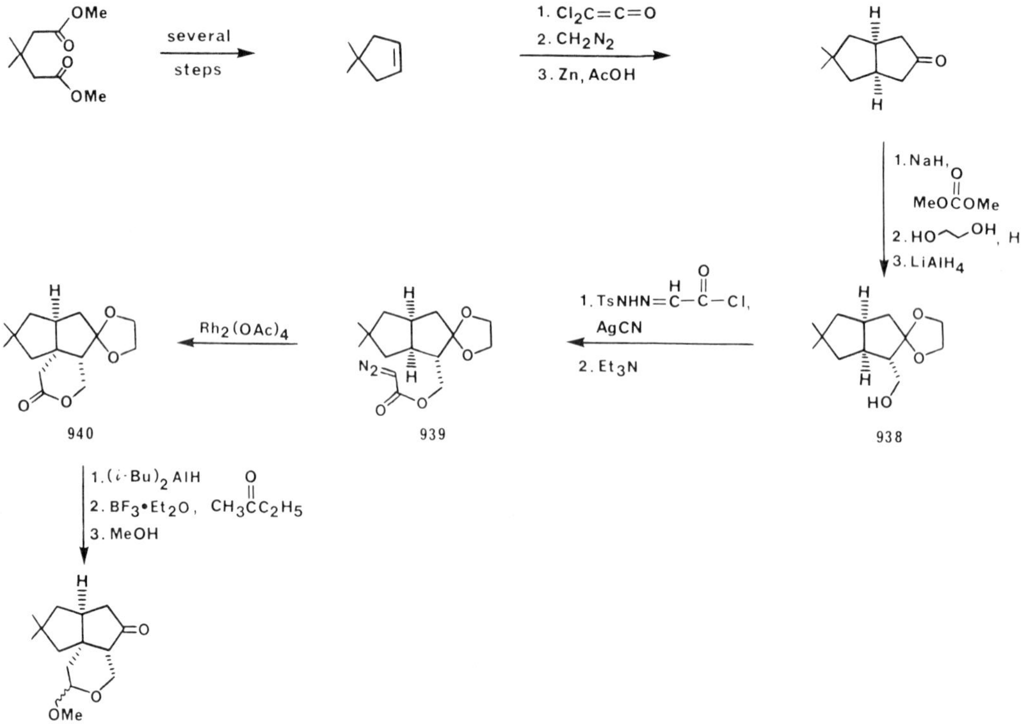

940   939   938

941

carbene into an unactivated carbon-hydrogen bond [415]. Acylation of bicyclic alcohol 938, obtained in several steps from dimethyl 3,3-dimethylglutarate [416], was achieved by reaction with glyoxalyl chloride tosylhydrazone in the presence of silver cyanide. Subsequent treatment with one equivalent of triethylamine afforded the diazo ester 939 in good yield. The key intramolecular carbene

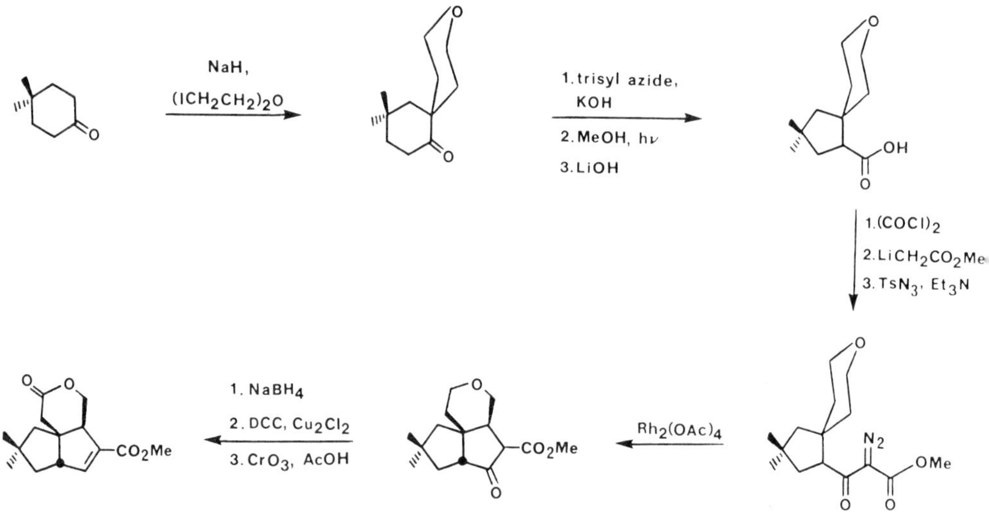

944   943   942

insertion reaction was carried out by slow addition of 939 to a suspension of rhodium(II) acetate in refluxing Freon, providing access to the lactone 940 (Scheme LXII). One of the epimeric acetals 941 has previously been converted to pentalenolactone E methyl ester (931) by Paquette [417]. Pentalenolactone F methyl ester (932) could be prepared from 931 using the method reported by Danishefsky for the synthesis of pentalenolactone (788) [418].

Employing a related rhodium-mediated intramolecular carbon-hydrogen insertion reaction, Taber and Schuchardt have recently reported a short synthesis of pentalenolactone E methyl ester [419]. The α-diazo-β-keto ester 942 obtained from 4,4-dimethylcyclohexanone was treated with catalytic rhodium(II) acetate in dichloromethane at room temperature, whereupon ring closure occurred to afford tricyclic ether 943. Lactone 944 has previously been converted to pentalenolactone E methyl ester (931) (Scheme LXIII) [417].

C Quadrone and Terrecyclic Acid

The sesquiterpene lactone quadrone (948), a fungal metabolite from *Aspergillus terreus* was shown to exhibit in vitro and in vivo cytotoxicity [420]. Due to its novel tetracyclic structure and interesting biological properties, quadrone has continued to be a challenging target for the refinement of synthetic strategy and development of synthetic methodology.

A short formal synthesis of (±)-quadrone reported by Yoshii and his coworkers employed the solvolytic rearrangement of an appropriately substituted

bicyclo[4.2.0]octanol as the key step [421]. Formolysis of 945 followed by basic hydrolysis and PCC oxidation afforded the rearranged bicyclic ketone 946. The remaining steps led to the Danishefsky intermediate 947 which has previously been converted to (±)-quadrone (948, Scheme LXIV) [422].

In the Schlessinger synthesis [423], the tricyclic skeleton of Danishefsky's intermediate 947 was constructed via an intramolecular Diels-Alder reaction of the α-methylene cyclopentanone derivative 949. Conversion to the *cis*-decalin was effected by allylic oxidation with chromium trioxide and 3,5-dimethylpyrazole followed by basic workup. Methylation, hydrogenation, and enol silane formation afforded 950. Aldol condensation of 1,4-diketone 951 gave enone 947, thus completing the formal synthesis of quadrone (948, Scheme LXV).

Burke and his coworkers have reported a modification to their original synthesis of (±)-quadrone [425] which employs a new approach to the δ-lactone. Deprotection of the earlier intermediate 952 followed by acid-catalyzed aldol condensation, elimination of acetic acid, and functional group manipulation afforded silyl ether 953. Ozonolytic cleavage of the double bond followed by reductive workup gave diol 954 which was easily converted to (±)-quadrone (948, Scheme LXVI) [425, 426].

The Vandewalle approach to ($\pm$)-quadrone [427] is very similar to the Schlessinger effort [423] in that it employs the same intramolecular Diels-Alder reaction to arrive at intermediate 950. The ensuing steps are slightly different, although the overall route to the Danishefsky intermediate 947 is longer.

As in the Danishefsky [422] and Helquist [428] approaches to ($\pm$)-quadrone, Pattenden has employed an intramolecular alkylation reaction to construct the six-membered carbocyclic ring [429]. Initially, Weiss-Cook condensation [430] of ethyl 4,5-dioxopentanoate with dimethyl acetone dicarboxylate afforded diquinane 955. Elaboration to bromo ester 956 set the stage for intramolecular alkylation and formation of tricyclic dione 957, which contains the carbocyclic ring system of ($\pm$)-quadrone (Scheme LXVII).

The first total synthesis of quadrone in chiral form, reported by Smith, allows the assignment of absolute stereochemistry [431]. The strategy is based upon the well established acid-catalyzed rearrangement of [4.3.2]propellanes. Photochem-

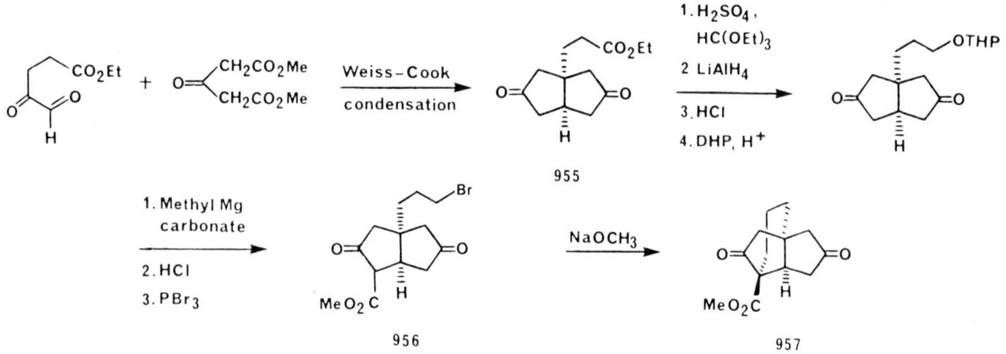

CO2Et + CH2CO2Me / CH2CO2Me

Weiss–Cook condensation

1. H2SO4, HC(OEt)3
2. LiAlH4
3. HCl
4. DHP, H+

955

1. Methyl Mg carbonate
2. HCl
3. PBr3

956

NaOCH3

957

MeO2C

958

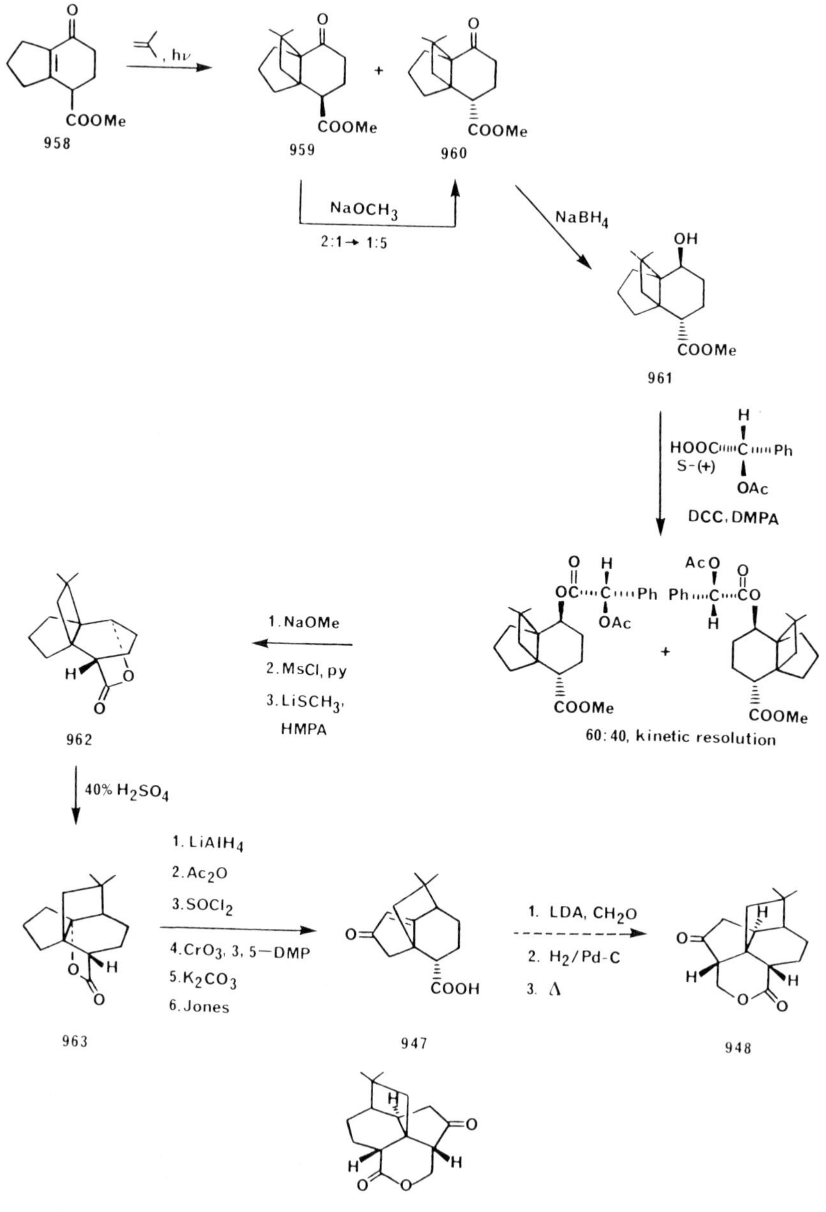

, hν

959 + 960

NaOCH3
2:1 → 1:5

NaBH4

961

COOMe

H
HOOC⋯C⋯Ph
S-(+)
OAc
DCC, DMPA

60:40, kinetic resolution

1. NaOMe
2. MsCl, py
3. LiSCH3, HMPA

962

40% H2SO4

963

1. LiAlH4
2. Ac2O
3. SOCl2
4. CrO3, 3,5–DMP
5. K2CO3
6. Jones

947

COOH

1. LDA, CH2O
2. H2/Pd–C
3. Δ

948

948*

ical [2 + 2] cycloaddition of isobutylene to bicyclic enone 958 followed by epimerization leads to a 1:5 mixture of propellanes 959 and 960. Stereospecific sodium borohydride reduction afforded the *anti*-propellane 961 whose resolution was achieved by separation of the diastereomeric O-acetylmandelic esters. Lactonization of the mesylate ester afforded the requisite [4.3.2]propellane 962. Under acidic conditions, 962 underwent rearrangement to lactone 963. Standard transformations then led to the customary Danishefsky intermediate 947 (Scheme LXVIII). The synthetic material was found to be enantiomeric with natural quadrone, thereby establishing the absolute configuration as 948*.

Cope rearrangement of a divinylcyclopropane derivative to introduce the 6-membered carbocyclic ring is the key step in the Piers approach [432] to quadrone (Scheme LXIX). Cyclopropanation of 964 by treatment with ethyl diazoacetate in the presence of rhodium(II) acetate provided an epimeric mixture of cyclopropanecarboxylate esters. After a reduction-oxidation sequence and base-promoted equilibration, a single aldehyde (965) was obtained and converted via standard reactions to enol silyl ether 966. Thermolysis of the latter intermedi-

ate and deprotection afforded the crystalline keto ketal 967. Reduction of the dimethylated derivative 968 with lithium diisobutyl-*n*-butylaluminum hydride avoided the problem of double bond reduction. Allylic alcohol 969 differs only in the ketal protecting group from an intermediate in the Burke approach [424–426] previously outlined.

A synthesis of (±)-descarboxyquadrone by Katiuchi and his coworkers [433] features acid-catalyzed rearrangement of the [4.3.2]propellanone 970 and is similar to the strategy adopted by Smith [431] in his chiral synthesis of quadrone. Since enone 971 has been converted to descarboxyquadrone [434] (972), a formal synthesis was realized.

In 1982 Sakai and coworkers reported the isolation of terrecyclic acid A from the fungus *Aspergillus terreus* [435]. It has an α-methylenecarbonyl arrangement reminiscent of a large number of known antitumor agents. A chiral synthesis of (−)-terrecyclic acid (762) from (+)-fenchone, which establishes its absolute configuration and that of quadrone, has recently been achieved (Scheme LXX) [436].

After the known conversion of fenchone to β-fenchocamphorane, Baeyer-Villiger oxidation, lactone cleavage, and oxidation gave the chiral cyclopentanone 973. Construction of the diquinane skeleton closely follows the Danishefsky approach [422] except that the methylene group is introduced earlier

974

1. t-BuOK
2. HO~OH, p-TsOH
3. LiAlH₄ → $LiAlH_4$
4. Ac₂O, py

1. LDA
2. MeO... Br / MeO₂C
3. H⁺

1. LDA, t-BuMe₂SiCl
2. 70–80°C
3. KF
4. MeI
5. p-TsOH
6. TiCl₄

1. (PhSe)₂, NaBH₄
2. HO~OH, p-TsOH
3. H₂O₂
4. Δ
5. TsCl
6. NaI

1. LiN(SiMe₃)₃
2. PrSLi
3. H₃PO₄

762

948*

in the sequence via an Ireland-Claisen rearrangement of the allyl acetate 974. Synthetic terrecyclic acid A was found to be the enantiomer of the natural material. Since its conversion to quadrone is known [435], the absolute configuration was established as 948* in agreement with Smith's result [431].

An approach to quadrone and terrecyclic acid by Iwata and his coworkers is interesting in that it employs a regioselective reductive ring opening of a cyclopropanecarboxylic acid followed by trapping of the enolate intermediate by a C-1 unit [437]. Specifically, reduction of 975 with lithium in liquid ammonia followed by quenching with phenylthiomethyl iodide and esterification gave the cyclopentanone 976. Deprotection of the alcohol and elimination of the phenylthio group by periodate oxidation furnished the enone 977, which has previously been transformed into both terrecyclic acid [436] and to quadrone [422].

1. PdCl₂, CuCl, O₂
2. HO~OH, p-TsOH
3. CH₂=CHMgBr, CuI•n-Bu₃P

1. HO~OH, p-TsOH
2. BH₃·THF
3. OH⁻, H₂O₂
4. 1 M HCl
5. DHP, p-TsOH
6. NaH, t-C₅H₁₁OH

1. Me₂S=CHCO₂Et
2. KOH

139

977   976   975

A conceptually very different approach to quadrone and terrecyclic acid has been described by Wender and Wolanin [438]. A novel nickel-mediated enone γ-alkylation gave access to diene 978. Intramolecular Diels-Alder cycloaddition

of triene 979 under Lewis acid catalysis gave only the *endo*-product 980. Saponification and halodecarboxylation afforded chloride 981, which underwent ring expansion on treatment with silver nitrate in dimethylformamide at 75 °C. Further elaboration led to terrecyclic acid A (762) (Scheme LXXI) which has previously been converted to quadrone [422].

D Ptychanolide

The sesquiterpenoid ptychanolide (987) isolated from *Ptychanthus striatus* in 1981 by Takeda et al. [439] possesses a diquinane carbon skeleton with four adjacent *cis*-located methyl groups. Dreiding and his coworkers have reported a synthesis of a stereoisomer of ptychanolide which features an α-alkynone cyclization to construct the bicyclo[3.3.0]octane system [440]. Passage of α-alkynone 982 through a quartz tube at 620 °C afforded the enone 983 (Scheme LXXII). The spiro-lactone moiety was incorporated via allylation of ester 984 followed by ozonolysis and basic hydrolysis to give a mixture of hydroxy lactones 985 (82:18). The unsaturated lactone 986, obtained via the thiophenyl ether, was epoxidized and after crystallization afforded one of the diastereoisomers of natural (+)-ptychanolide (987).

141

E Albene

The trisnorsesquiterpene (−)-albene, first isolated in 1962 from *Petasites albus* [441], is now known to be the 1S,2S,6S,7R enantiomer of 2-*endo*,6-*endo*-dimethyltricyclo[5.2.1.0^{2,6}]dec-3-ene (988) [442].

988

However, the correct structure and absolute stereochemistry of (−)-albene have been difficult to assign. The initial degradation work in 1972 [443] was wrongly interpreted and it was not until 1977 that albene was shown to possess endo methyl groups [444]. Baldwin and coworkers have subsequently provided an explanation for this misinterpretation and have unambigously proven the absolute configuration of (−)-albene [442, 445, 446].

Although the Kreiser group did establish the structure of (−)-albene, they in fact assigned the wrong absolute stereochemistry in their synthetic work [447]. Their route from (+)-camphenilone employed a rearrangement of chloro alcohol 989 in formic acid at reflux to afford the tricyclic ketone 990.

989    $\xrightarrow[\text{100}^{\circ}\text{C}]{\text{HCOOH}}$ 990

1. CF$_3$CO$_3$H
2. 25% KOH

+

Pb(OAc)$_4$, py
C$_6$H$_6$, Δ

Baldwin and Barden, in an extensive investigation of this reaction [445], showed that it in fact occurs with considerable racemization and does not depend on an endo-3,2-methyl shift in the substituted 2-norbornyl cationic intermediate.

In recent years, three independent syntheses of (±)-albene have been reported. The first, due to Baldwin [448], features the Diels-Alder reaction of 2,3-dimethylmaleic anhydride and cyclopentadiene (Scheme LXXIII). Separa-

tion of the hydrogenated exo and endo adducts, followed by reaction of the former with methyllithium, protection, and subsequent reduction with diisobutylaluminum hydride gave 991. Rearrangement to enone 992 was effected by exposure to methanolic potassium hydroxide in the presence of 3 Å molecular sieves followed by acidification. (±)-Albene was then synthesized from cyclopentenone 992 via a previously established route [449, 450].

The second synthesis, due to the Trost group [451], utilizes a palladium-catalyzed cycloaddition of 2-[(trimethylsilyl)methyl]-3-acetoxy-1-propene to 2,3-dicarbomethoxynorbornene as the key step. This efficient five-step synthesis of (±)-albene is summarized in Scheme LXXIV.

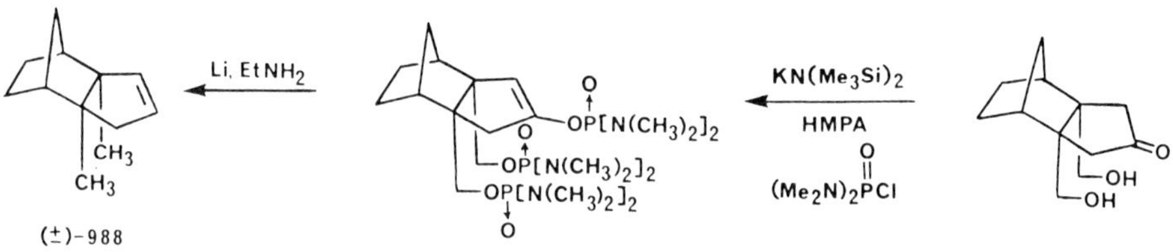

(±)-988

In the most recent synthesis of (±)-albene, Dreiding and coworkers chose to employ a highly regiospecific thermal α-alkynone cyclization to introduce a new quaternary carbon atom at an unactivated position [452]. The Diels-Alder adduct 993 of tiglyl chloride and cyclopentadiene was elaborated to the α-alkynone 994. Thermolysis at 580 °C gave enone 995, and this major product was easily converted to (±)-albene (Scheme LXXV).

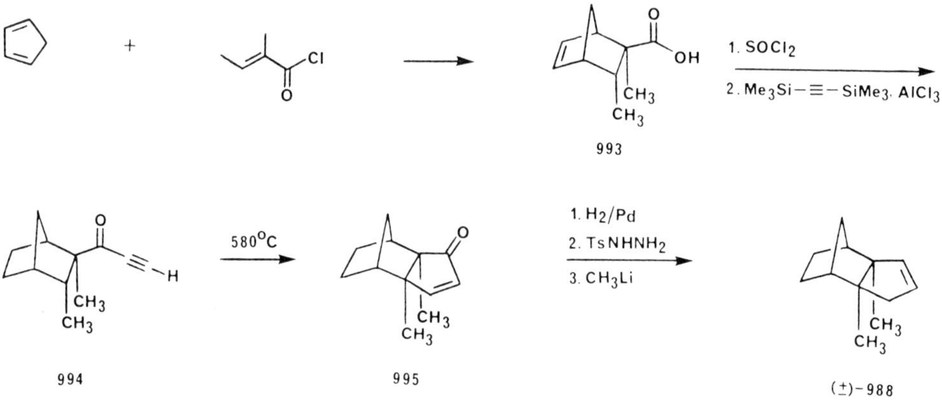

F Carbaprostacyclins

Due to the potent antiaggregatory and vasodilating properties of prostacyclin (996) [453], coupled with its inherent instability [454], considerable efforts have been expended to develop stable and therapeutically useful analogues [455]. In the past few years, activity in this field has been particularly intense, resulting in the discovery of several new biologically active mimics for prostacyclin and in the development of new synthetic methods to achieve their synthesis.

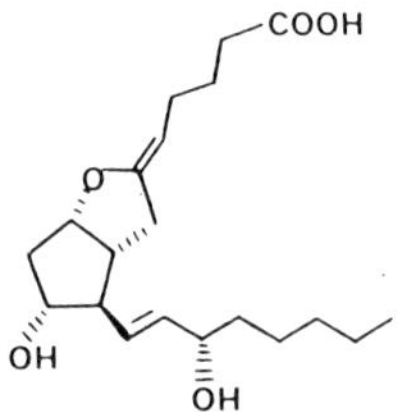

996

The Ikegami group has made a significant contribution to the synthesis of several carbon analogues of prostacyclin (PGI$_2$). Thus, 9(O)-methano-$\triangle^{6(9\alpha)}$-PGI$_1$ (997), first synthesized by utilizing an intramolecular pinacolic coupling reaction [456], was shown to be more potent than the previously reported analogue 9(O)-methanoprostacyclin (carbacyclin, 998) in inhibition of platelet aggregation.

997

998

The synthesis of 997, summarized in Scheme LXXVI, began with the optically active ketone 999. Methylenation, hydroboration, and treatment with alkaline hydrogen peroxide led to primary alcohol 1000. Iodoetherification and heating with DBU in toluene gave enol ether 1001. After treatment with aqueous acetic acid and oxidation to keto aldehyde 1002, the key pinacol coupling reaction was effected by exposure to titanium(IV) chloride and zinc to afford the bicyclic diols 1003 as a diastereoisomeric mixture. With the bicyclo[3.3.0]octane skeleton elaborated, the desired endo olefinic bond was introduced via epoxide 1004. Finally, hydrolysis and decarboxylation of 1005 yielded the new prostacyclin analogue 997.

145

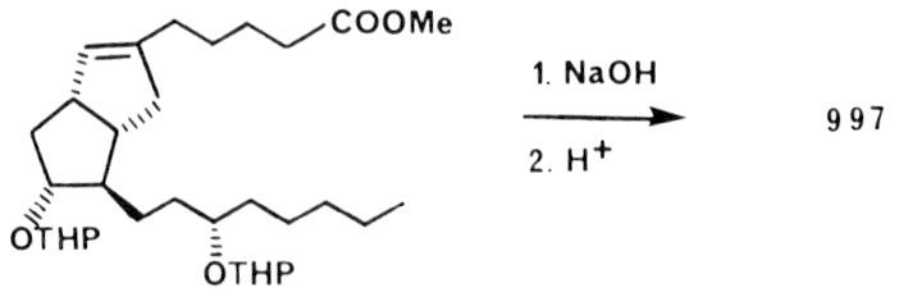

COOMe

1. NaOH
   → 997
2. H⁺

1005

An improved synthesis of 997 reported by the same group [457] features the regiocontrolled transformation of enone 1006 to the endo-olefin 1005 shown in Scheme LXXVII. Thus, Michael addition of the trimethylsilyl anion to 1006 followed by reduction and subsequent treatment with trifluoromethanesulfonic anhydride, pyridine and 4-dimethylaminopyridine provided the olefin 1005 previously obtained.

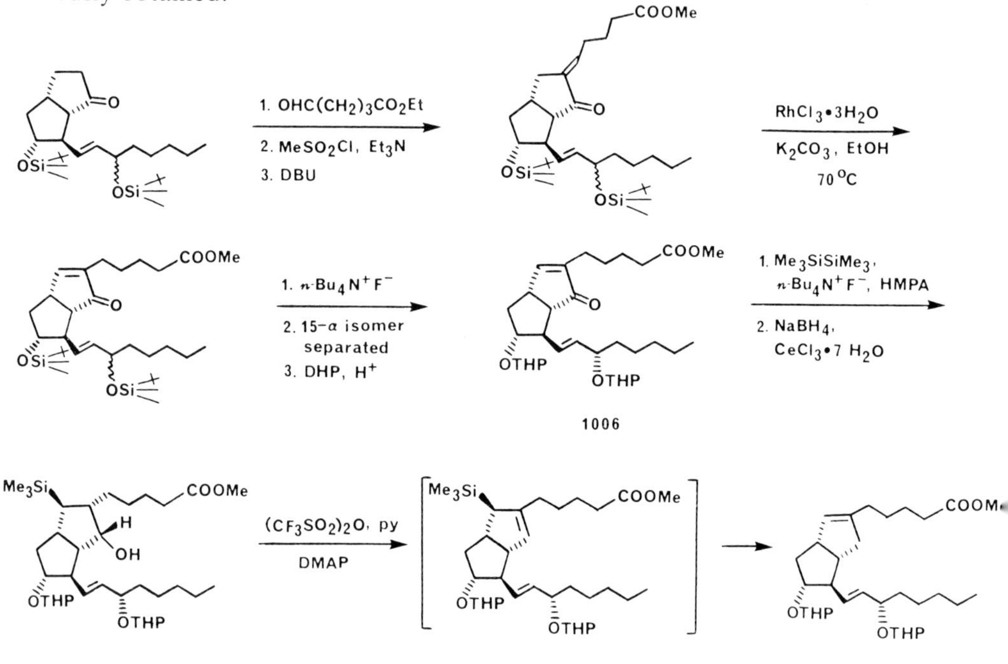

1. OHC(CH₂)₃CO₂Et
2. MeSO₂Cl, Et₃N
3. DBU

RhCl₃•3H₂O
K₂CO₃, EtOH
70 °C

1. n-Bu₄N⁺F⁻
2. 15-α isomer separated
3. DHP, H⁺

1006

1. Me₃SiSiMe₃, n-Bu₄N⁺F⁻, HMPA
2. NaBH₄, CeCl₃•7 H₂O

(CF₃SO₂)₂O, py
DMAP

1005

Recognizing that the chemically unstable △⁶-prostaglandin I₁ (△⁶-PGI₁, 1007) is highly active in inhibiting platelet aggregation [458], the Ikegami group developed two routes to 9(O)-methano-△⁶-prostaglandin I (1008) [459] in order to evaluate its biological properties.

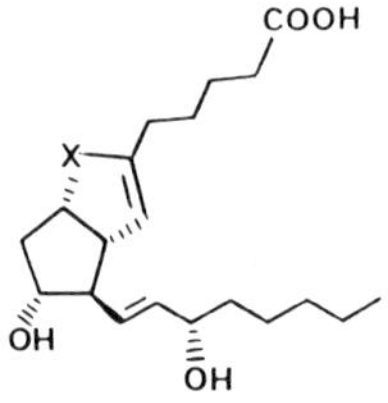

COOH

1007, X = O
1008, X = CH₂

The first route (Scheme LXXVIII), which suffers from a low overall yield, features decarboxylative elimination of β-mesyloxy ester 1009. In the second more efficient route (Scheme LXXVIII), enone 1010 was formed via an aldol condensation and hydrogenated to afford ketone 1011 with regio- and stereochemical control. L-Selectride reduction furnished endo-alcohol 1012 stereospecifically. The double bond was introduced by elimination of the corresponding mesylate with DBU. Deprotection followed by ester hydrolysis led to 1008 (as a mixture of C-15 isomers).

1013   1014

Two additional analogues 1013 and 1014 synthesized by Ikegami [461] showed weak inhibitory activity in rabbit platelet aggregation by adenosine diphosphate. Both compounds were obtained from the same synthetic intermediate 1011 [459].

In the synthesis of 1014 (Scheme LXXIX), exo alcohol 1015, obtained as the minor product from the sodium borohydride reduction of 1011 (along with the endo alcohol 1012, 1:3 ratio) [459], could be converted to the phenoxythiocarbonyl derivative 1016 according to the method of Robins [462]. Reduction of 1016 with tributyltin hydride followed by deprotection and hydrolysis afforded a separable mixture of 1014 and its C-15 stereoisomer. The same starting ketone 1011, after base-catalyzed epimerization and reduction gave hydroxy ester 1017 as the only product. Removal of the alcohol moiety was effected as before via the corresponding phenoxythiocarbonyl derivative 1018 (Scheme LXXIX).

1011′

NaBH₄

1. NaOMe, MeOH
2. NaBH₄

1015   +   1012   1017

(1 : 3)

S
‖
PhO–C–Cl, DMAP

1016

1018

1. $n\text{-}Bu_3SnH$
2. $n\text{-}Bu_4N^+F^-$
3. KOH

1014

+

+

1013

In an interesting synthesis of $(+)$-9(O)-methano-$\triangle^{6(9\alpha)}$-PGI$_1$ (997), Shibasaki employed a thermal ene reaction to construct the bicyclo[3.3.0]octane framework [463]. With optically active Corey lactone as starting material, elaboration of prostanoid aldehyde 1020 via diene 1019 was achieved by a number of standard transformations (Scheme LXXX). When 1020 was heated at 180 °C in toluene, the two ene products 1021 and 1022 were obtained in 87% combined yield (3:5 ratio). The mixture was hydrogenated and the endocyclic double bond introduced via DBU-promoted elimination of the corresponding mesylate derivatives. The remaining sidechain of 9(O)-methano-$\triangle$-PGI$_1$ was incorporated in the usual way by a Horner-Wittig coupling of aldehyde 1023 with sodium dimethyl(2-oxoheptyl)phosphonate followed by conventional functional group manipulation.

The Shibasaki approach to a more practical synthesis of $(+)$-997 [464], suitable for the preparation of multigram quantities, made use of an intramolecular aldol condensation reaction for the construction of bicyclo[3.3.0]octane

1. $(i\text{-}Bu)_2AlH$
2. $t\text{-}BuOK$, $Br^-PhP_3^+(CH_2)_3CO_2H$
3. CH_2N_2
4. PCC, NaOAc
5. Zn, CH_2Br_2, $TiCl_4$

1019

1. Me_2AlCl
2. $t\text{-}BuMe_2SiCl$, imidazole, DMF
3. 9-BBN
4. OH^-, H_2O_2
5. Collins

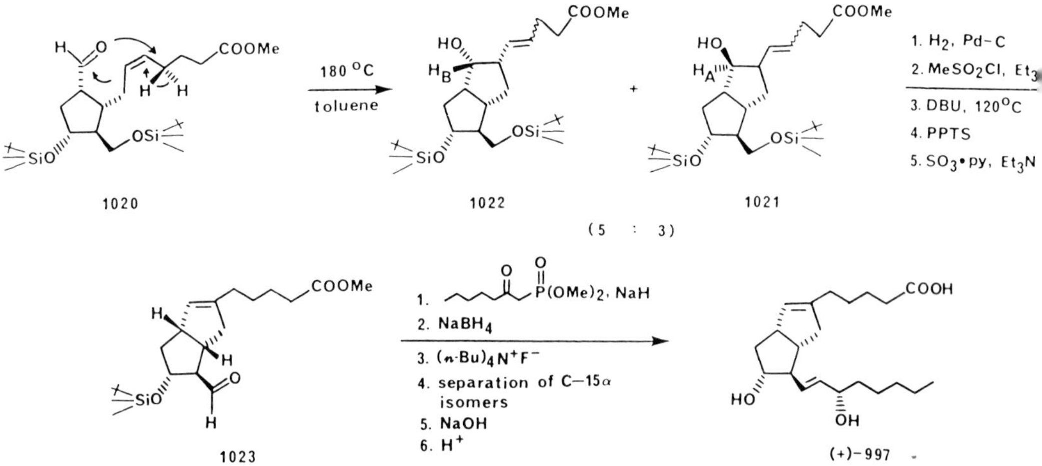
COOMe
180 °C
toluene
HO
H_B
COOMe
HO
H_A
COOMe
1. H_2, Pd–C
2. $MeSO_2Cl$, Et_3
3. DBU, 120°C
4. PPTS
5. SO_3•py, Et_3N
1020
1022
1021
(5 : 3)
COOMe
1. $P(OMe)_2$, NaH
2. $NaBH_4$
3. $(n\text{-}Bu)_4N^+F^-$
4. separation of C–15α isomers
5. NaOH
6. H^+
COOH
HO
OH
1023
(+)-997

derivative 1025. The precursor dialdehyde 1024 was obtained from optically pure Corey lactone by a similar approach to that described in Scheme LXXX [463]. The cis:trans (1:2) mixture of dienes 1026 obtained after a Wittig reaction with the α,β-unsaturated aldehyde 1025 followed by diazomethane esterification was regioselectively hydrogenated to afford the desired bicyclo[3.3.0]octene

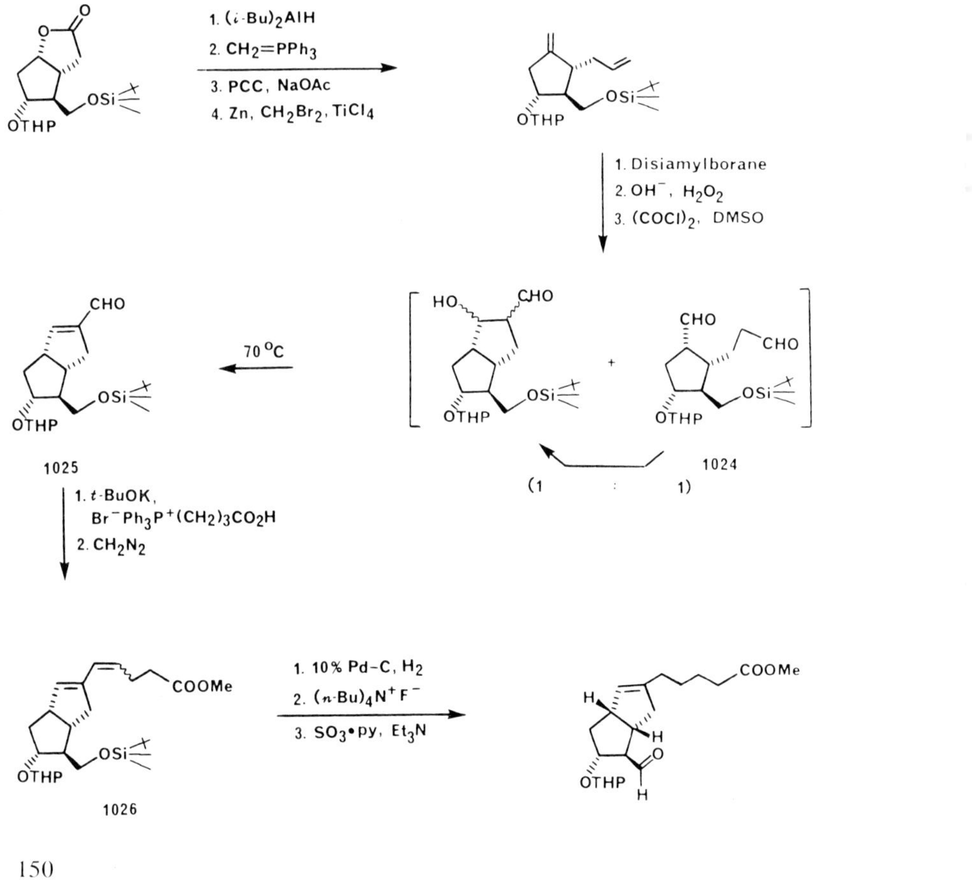
1. $(i\text{-}Bu)_2AlH$
2. $CH_2{=}PPh_3$
3. PCC, NaOAc
4. Zn, CH_2Br_2, $TiCl_4$
OTHP
1. Disiamylborane
2. OH^-, H_2O_2
3. $(COCl)_2$, DMSO
CHO
70 °C
HO
CHO
CHO
CHO
1025
1024
(1 : 1)
1. t-BuOK, $Br^-Ph_3P^+(CH_2)_3CO_2H$
2. CH_2N_2
COOMe
1. 10% Pd–C, H_2
2. $(n\text{-}Bu)_4N^+F^-$
3. SO_3•py, Et_3N
COOMe
1026

framework. Elaboration to (+)-997 was realized in the customary manner (Scheme LXXXI).

As a continuation of their efforts to improve the synthesis of (+)-997, Ikegami and Shibasaki have reported an additional route in which the regiospecific transformation of diene 1027 into diol 1028 was achieved [465]. Thus 1027, obtained from the Corey lactone, was reacted with thexylborane and the resulting boracycle oxidized to afford the isomeric diols 1028a and 1028b (ca 7:1). Without separation, these diols were oxidized and converted to the desired endo olefin 1029 by means of the pinacol coupling strategy previously reported. Modification of the ω-sidechain was achieved by variation of the phosphonate reagent used in the Horner-Emmons reaction with aldehyde 1029 (Scheme LXXXII). Analogues 1030 and 1031 were found to be slightly more potent than 997 in the inhibition of platelet aggregation.

In another route to (+)-997, Ikegami and his coworkers [466] introduced the endocyclic double bond via an allylic rearrangement reaction. Desulfurization of the rearranged product 1033 obtained from allylic alcohol 1032 furnished 1034 which has already been converted to 997 [456, 457, 465].

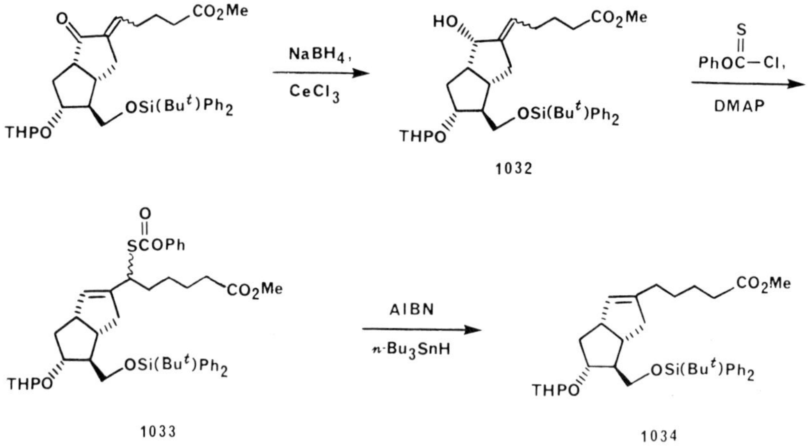

1032

1033 → 1034

One of the major problems associated with most of the synthetic approaches to carbacyclin 998 has been the failure to introduce the 5(*E*)-trisubstituted olefinic center stereospecifically. Shibasaki has solved this problem by utilizing the stereospecific 1,4-hydrogenation of a 1,3-diene [467]. Toward this end, 1036 was synthesized from the Corey lactone by a previously reported strategy involving an intramolecular thermal ene reaction within the allylic alcohol 1035 [463]. Stereospecific hydrogenation of *trans*-1036 using (methylbenzoate)Cr(CO)$_3$ as a catalyst in acetone solution under 70 atm of hydrogen pressure at 120 °C provided 1037 in almost quantitative yield. Under the same conditions, the (*Z*)-rich 1,3-diene 1038 gave only the desired product 1039. Both 1037 and 1039 could easily be converted to carbacyclin 998 by previously reported methods [468].

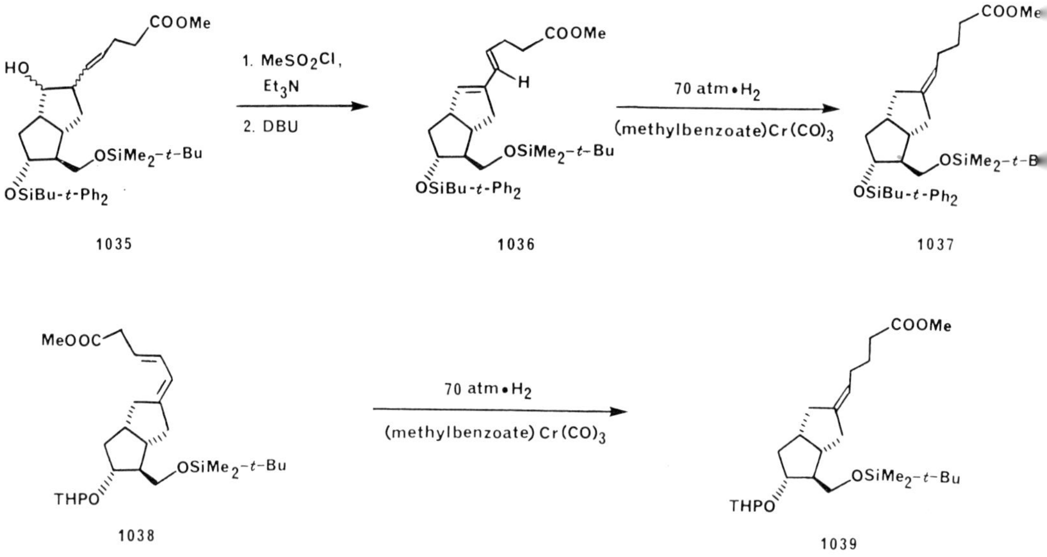

1035 → 1036 → 1037

1038 → 1039

The structures and reaction schemes are shown.

1025

naphthalene·Cr(CO)$_3$ or
(methylbenzoate)Cr(CO)$_3$

H$_2$, 120°C

1. (i-Bu)$_2$AlH
or NaBH$_4$
2. CBr$_4$, PPh$_3$
3. KCN, MeCN
18-crown-6

1. LDA
2. OHC(CH$_2$)$_2$CO$_2$Me

MeSO$_2$Cl,
Et$_3$N

(83%)

(10%)

1. n-Bu$_4$N$^+$F$^-$
2. SO$_3$·py, DMSO, Et$_3$N
3. (MeO)$_2$POCHNaCOR
4. NaBH$_4$
5. AcOH, H$_2$O, THF
6. NaOH

1040a, R = C$_5$H$_{11}$
b, R = CH(Me)C$_4$H$_9$

The cyanocarbacyclins 1040a and 1040b have also been synthesized by a protocol involving 1,4-hydrogenation of a conjugated diene catalyzed by an arene·Cr(CO)$_3$ complex [469]. Their syntheses from Corey's lactone are summarized in Scheme LXXXIII [464].

Two additional routes to 9(O)-methano-$\triangle^{6(9\alpha)}$-PGI$_2$ (997) due to Shibasaki [470] began from furfural in an effort to avoid dependence on the availability of the Corey lactone. Allyl ketone 1041 [471] was converted to the α,β-unsaturated aldehyde intermediate 1042 by standard methodology. In the first route, diene 1043 was regioselectively hydrogenated using a modified Wilkinson catalyst obtainable from chlorodicyclooctene rhodium(I) and dipiperidylphosphine to afford olefin 1040 that was easily converted to 997. In the second route, the α-sidechain was incorporated via a regioselective organocopper coupling involving allylic acetate 1045 to afford mainly the α-product 1046. This intermediate was further elaborated to 997 (Scheme LXXXIV).

Aristoff and his coworkers have synthesized a series of 9-substituted carbacyclin analogues found to possess potent platelet antiaggregatory activity [472]. Their strategy was based upon the nickel-catalyzed conjugate addition of a

variety of alkynyl aluminates to the known enone 1047 [473]. The modified Schwartz procedure [474] was utilized in the synthesis of 1048 a, b, c. The cyano ketone 1050 could also be easily prepared from enone 1047 by reaction with acetone cyanohydrin, potassium cyanide, and 18-crown-6 in toluene. The α-sidechain could be incorporated without difficulty to afford compounds 1049 a, b, c (Scheme LXXXV) but problems of β-elimination were experienced in the same transformation of nitrile 1050. This complication was resolved by a two-step procedure involving addition of a carboxylic acid dianion followed by decarboxylative dehydration (Scheme LXXXVI). 9-Cyanocarbacyclin (1051), 9-ethynyl-carbacyclin (1049 c), and 9-(1'-propynyl)carbacyclin (1049 b) were all found to be more active than carbacyclin 998 at inhibiting platelet aggregation. The acetylenic analogues 1049 b and 1049 c were approximately twice as potent as prostacyclin 996 in the same test.

The synthetic prostacyclin analogue CG 4305 (1052) possesses similar biological activity to prostacyclin 996. Flohé and coworkers have reported the X-ray structure of this analogue in order to gain insight into its conformational properties [475]. The results indicate the *m*-carboxyphenyl moiety to be essentially

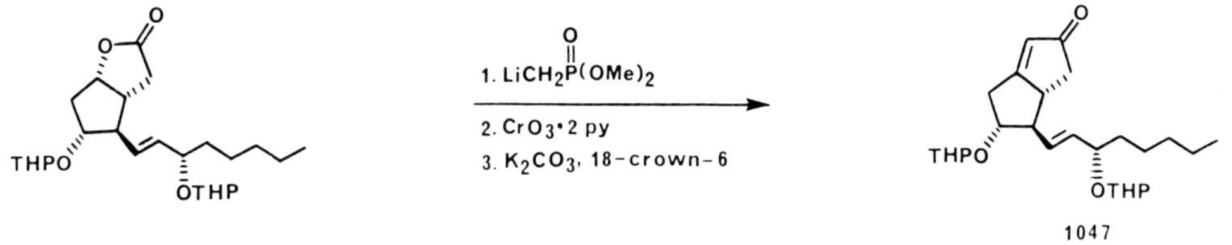
1. LiCH2P(OMe)2 (O)
2. CrO3·2 py
3. K2CO3, 18-crown-6
THPO
OTHP
1047

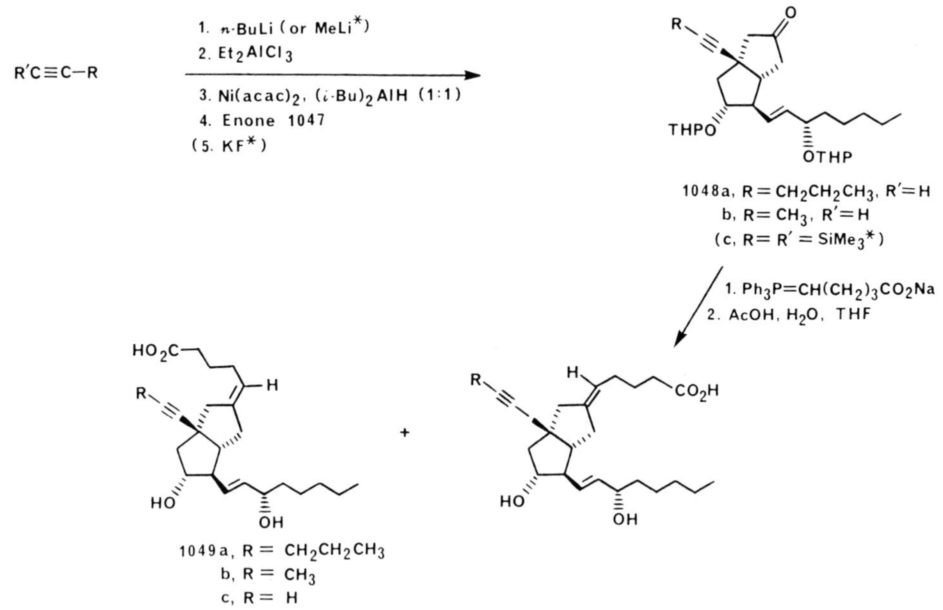
R'C≡C—R
1. n-BuLi (or MeLi*)
2. Et2AlCl3
3. Ni(acac)2, (i-Bu)2AlH (1:1)
4. Enone 1047
(5. KF*)
THPO
OTHP
1048a, R = CH2CH2CH3, R' = H
b, R = CH3, R' = H
(c, R = R' = SiMe3*)
1. Ph3P=CH(CH2)3CO2Na
2. AcOH, H2O, THF
HO2C
CO2H
HO
OH
+
1049a, R = CH2CH2CH3
b, R = CH3
c, R = H

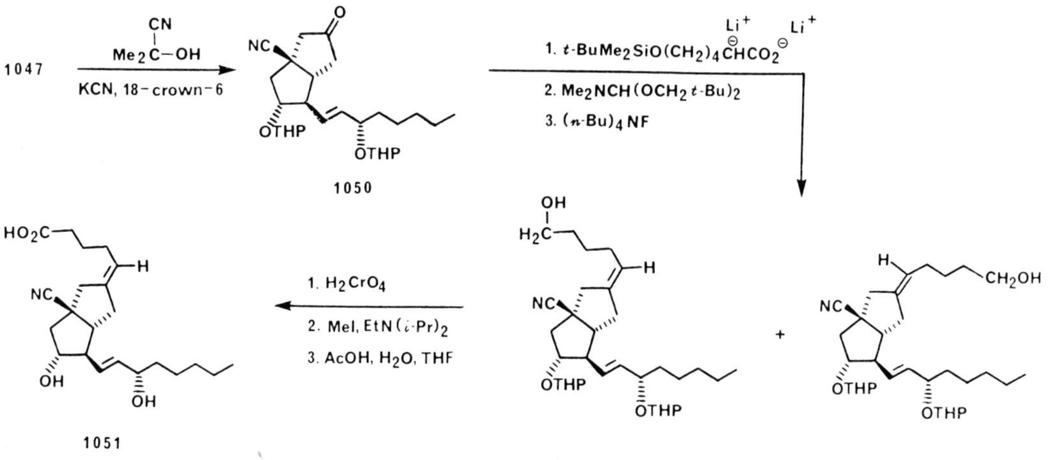
1047
CN
Me2C—OH
KCN, 18-crown-6
NC
OTHP
1050
1. t-BuMe2SiO(CH2)4CHCO2 Li+ Li+
2. Me2NCH(OCH2 t-Bu)2
3. (n-Bu)4NF
OH
H2C
CH2OH
+
1. H2CrO4
2. MeI, EtN(i-Pr)2
3. AcOH, H2O, THF
HO2C
OH
1051

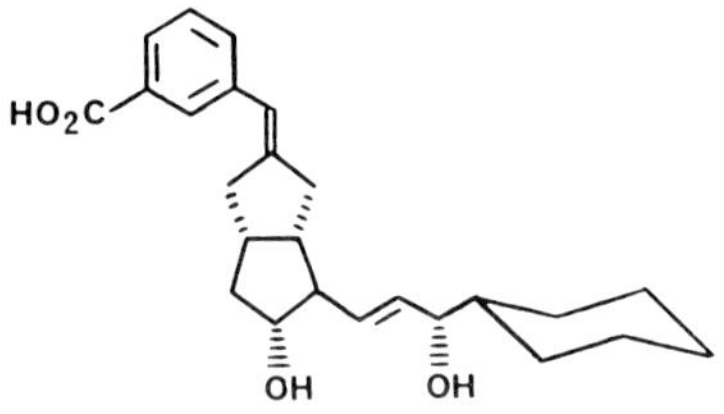

C G 4035

1052

planar and for there to be considerable conjugation with the exocyclic double bond leading to less conformational freedom than the natural product.

The same group, in an extension of their studies on the carboxyphenyl modified prostacyclin family, has synthesized a variety of new analogues enabling them to discover useful structure-activity relationships [476].

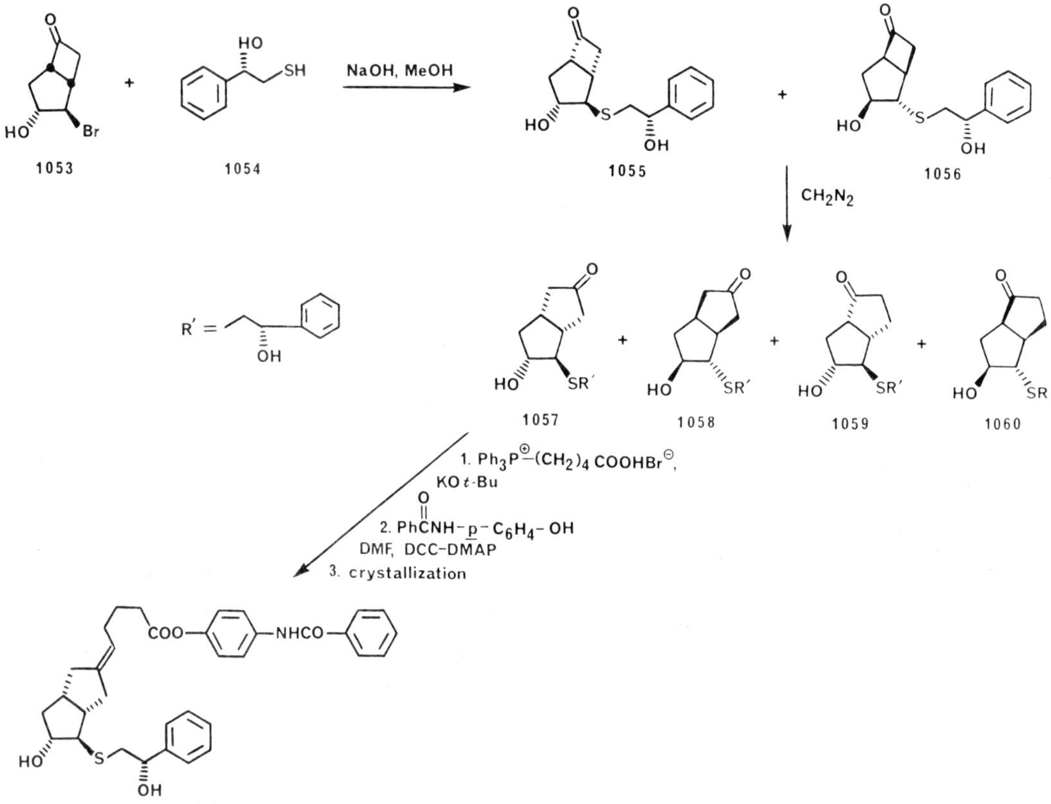

A short synthesis of the optically active carbacyclin analogue 1061 starting from cyclopentadiene was achieved by Riefling [477]. This compound was found to be highly active in lowering the blood pressure of renal hypertensive dogs. The known bromohydrin 1053 [478, 479] was reacted with thiol 1054 (obtained from L-(+)-mandelic acid) [480] in the presence of sodium hydroxide to afford a 1:1 mixture of 1055 and 1056 (Scheme LXXXVII). Since difficulty was experienced

with their separation, the mixture was ring expanded with diazomethane to afford the easily separable ketones 1057–1060. Compound 1057 (23%) was further elaborated to the carbacyclin 1061.

A more efficient approach [481] to optically active 13-thiacarbacyclins reported by Riefling also begins with bromohydrin 1053 [477]. Resolution was achieved by reaction of bicyclic ketone 1062 with (−)-ephedrine followed by separation of the diastereoisomeric derivatives 1063. Ring expansion of epoxide 1064 using lithium iodide in tetrahydrofuran at room temperature gave the bicyclo[3.3.0]octyl ketone 1065, easily converted in turn to the 13-thiacarbacyclin 1066.

In his synthesis of the stable prostacyclin analogue 997, Kojima chose to introduce the endo-olefinic moiety via cyclopropyl keto ester 1068 obtained in several steps from bicyclic monoacetal 1067 [482]. Cleavage of the cyclopropane ring with formic acid in the presence of concentrated sulphuric acid followed by reduction and protection of the hydroxyl group gave the β-formate 1069. Methanolysis of the formate to the corresponding alcohol, mesylate formation and base-promoted elimination afforded bicyclo[3.3.0]octene 1070, easily transformed in turn to (±)-997 (Scheme LXXXVIII).

For the synthesis of the isomeric compounds 1073 and 1074, the Kojima group began with a Wittig reaction of the known optically active aldehyde 1071 with the β-oxido ylide reagent 1072 (Scheme LXXXIX) [483]. In a preliminary test, 1073 was found to be a potent inhibitor of platelet aggregation while 1074 was appreciably less active.

The same group has reported an interesting stereoselective method for the introduction of the 15α-hydroxy group during syntheses of (±)-6,9α-methano-

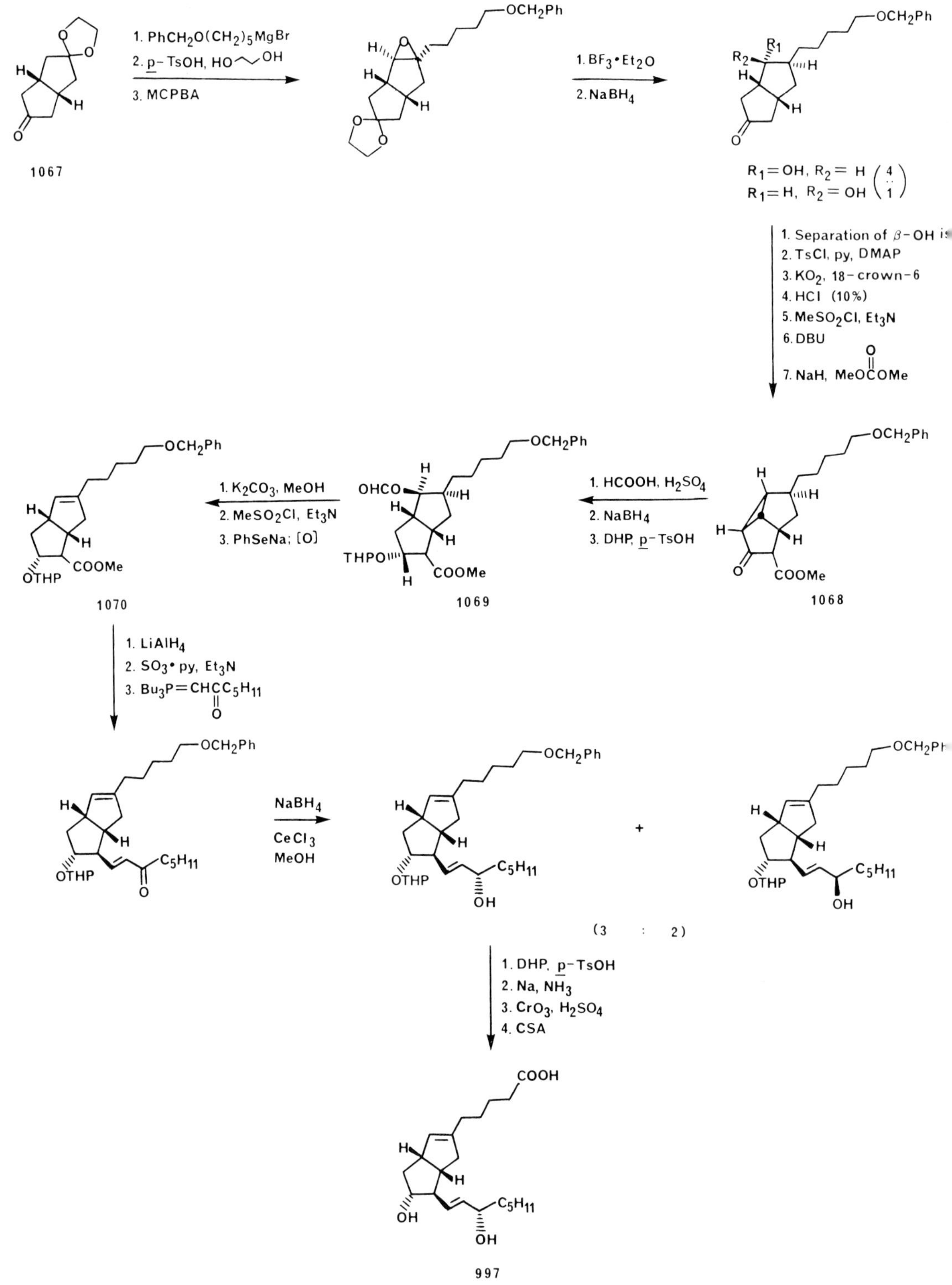

1067

1. PhCH₂O(CH₂)₅MgBr
2. p-TsOH, HO⌒OH
3. MCPBA

1. BF₃·Et₂O
2. NaBH₄

R₁=OH, R₂= H (4)
R₁=H, R₂= OH (1)

1. Separation of β-OH is
2. TsCl, py, DMAP
3. KO₂, 18-crown-6
4. HCl (10%)
5. MeSO₂Cl, Et₃N
6. DBU
7. NaH, MeOCOMe

1070

1. K₂CO₃, MeOH
2. MeSO₂Cl, Et₃N
3. PhSeNa; [O]

1069

1. HCOOH, H₂SO₄
2. NaBH₄
3. DHP, p-TsOH

1068

1. LiAlH₄
2. SO₃·py, Et₃N
3. Bu₃P=CHCC₅H₁₁

NaBH₄
CeCl₃
MeOH

+

(3 : 2)

1. DHP, p-TsOH
2. Na, NH₃
3. CrO₃, H₂SO₄
4. CSA

COOH

C₅H₁₁

OH   OH

997

prostaglandin I$_3$ (1073) and (±)-5,6-dihydro-6,9α-methano-6β-prostaglandin I$_3$ (1079). The 13(S)-configuration of aldehydes 1076 and 1078 was generated in the addition of phenylsulfenyl chloride to the less hindered side of enol ethers 1075 and 1077. After homologation and oxidation, the resulting allylic sulfoxides

underwent [2,3]-sigmatropic rearrangement to the desired 15α alcohol deriva-tives. The overall syntheses of 1073 and 1079 are summarized in Schemes XC and XCI, respectively.

A recent chiral synthesis of (+)-9(O)-methanoprostacyclin (998) [485] and its isomer 1083 reported by the Kojima group began with the optical resolution of the racemic acid 1080 [486–488]. This was accomplished by recrystallization of either the d(+)- or l(−)-α-methylbenzylamine salt from chloroform, followed by diazomethane esterification. The resulting two bicyclic esters 1081 and 1082 were transformed into 998 and 1083 using standard methodology.

Introduction of the upper sidechain in many of the carbacyclins has been achieved by Wittig chemistry. The process invariably leads to mixtures of the desired biologically active E-isomers and the much less active Z-isomers. Vor-brüggen and Bennua have reported a partial solution to this problem by establish-ing a route in which the Z-isomer may be easily recycled [489]. In their work, the undesired Z-isomer was oxidized under Lemieux-Johnson conditions to regener-ate the bicyclic ketone (1085) from which it was derived.

The Bestmann method [490] for the facile ring closure of ω-keto acids with triphenyl(phenyliminovinylidene)phosphorane 1087 constitutes the key step in

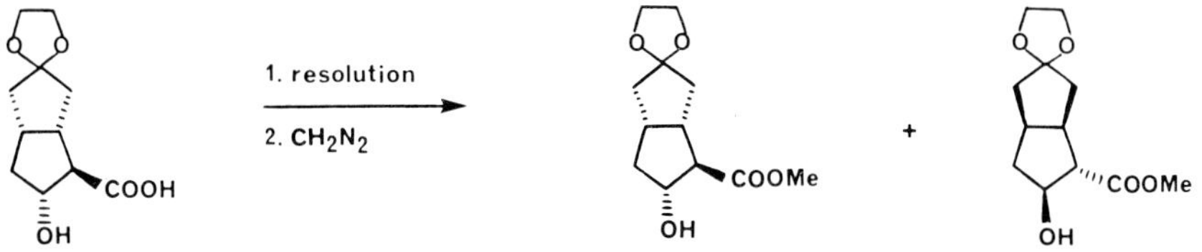

1080

1081

1082

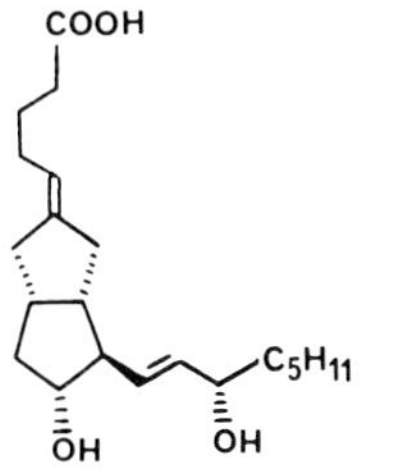

998

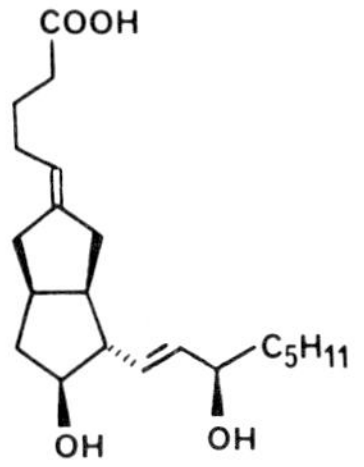

1083

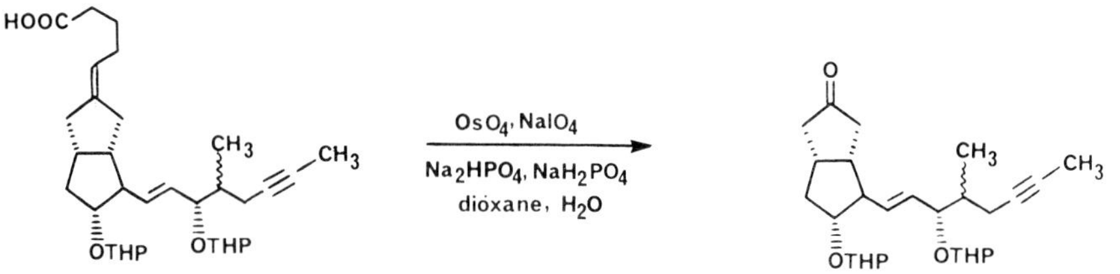

1084

1085

the chiral synthesis of unsaturated bicyclo[3.3.0]octan-3-one derivatives by Vorbrüggen and coworkers [491]. This strategy is similar to the intramolecular
Horner-Wittig cyclization previously reported by Aristoff [473]. The ω-keto acids
1086 a—d were easily prepared from the Corey lactone. Upon heating for 3 hours
with 1087 in boiling ethyl acetate, the ylide derivatives 1088 a—d were obtained.
Cyclization and elimination of N-phenylurethane gave the bicyclic enones
1089 a—d in 22—54% overall yield.

The key step in the synthesis of the stable enone analogue 1093 was oxidative
ring contraction of 1091 to the bicyclo[3.3.0]octane carboxylic acid 1092 [492].

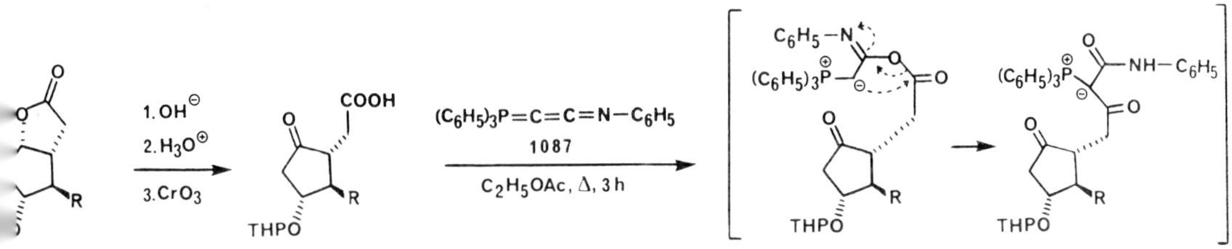

1086a–d

1087

1088a–d

toluene / C$_2$H$_5$OH
Δ, 4–12h

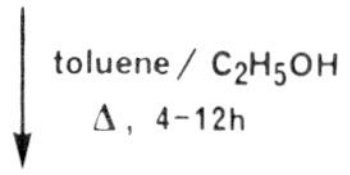

	R	Yield (%) of 1089
a	$-CH_2-O-CH_2-\bigcirc$	40
b	$-CH_2-O-Si(CH_3)_2C_4H_9-t$	22
c	(alkene chain with CH_3, CH_3, $\overset{\cdot}{O}THP$)	50
d	(alkene chain with CH_3, CH_3, alkyne, $\overset{\cdot}{O}THP$)	54

1089a–d

Coffee and coworkers commenced their synthesis with the hydrindanone 1090, reduction, acetylation, and acid hydrolysis of which produced 1091. Ring contraction to 1092 was effected by treatment with thallic nitrate in acetic acid. The two routes for attachment of the ω-sidechain are summarized in Scheme XCII. Preliminary biological evaluation has indicated 1093 to be approximately 1000 times less active than prostacyclin sodium salt in inhibiting collagen-induced

1. $(i\text{-Bu})_2$AlH
2. Ac_2O, py
3. HCl, MeOH

1090

1091

$Tl(NO_3)_3$
AcOH, Δ

1. HCl, MeOH
2. PCC
3. HO⌒OH, H^+
4. L–Selectride
5. PCC

1092

$R' = $ cyclohexyl

1. $(MeO)_2P(O)CH_2COR'$, NaH
2. L–Selectride
3. separation of $15\alpha,\beta$–alcohols
4. AcOH–H_2O
5. $ClSiMe_2 t$-Bu

1. B_2H_6
2. PCC
3. $(MeO)_2P(O)\overset{\ominus}{C}HCOR'$
4. K–Selectride
5. $ClSiMe_2 t$-Bu
6. K_2CO_3
7. PCC

$OSiMe_2 t$-Bu   + $OSiMe_2 t$-Bu

1. LBTMSA,
   MeO$_2$C(CH$_2$)$_3$CHO
2. DBU
3. AcOH–THF–H$_2$O
4. LiOH

CO$_2$H

ÖH

1093

aggregration, but to compare favorably with non-prostaglandin inhibitors of platelet aggregation.

The Mori group became interested in a single chiral synthesis of carbacyclin 998 [493], since most of those reported to date are racemic in nature. Kinetic resolution of the β-keto ester 1094 by modified yeast reduction led to (+)-1094 in 92—94%.

CO$_2$Et

(±)-1094

Saccharomyces bailii KI 0116

phosphate buffer
pH = 7

CO$_2$Et

(+)-1094

+

EtO$_2$C

HO (S)

(+)-1095

The first synthesis of a new prostacyclin analogue 4,4,5,5-tetrahydro-9(O)-methano-△-PGI$_1$ (1098) developed by the Shibasaki group [495] employs the regiocontrolled cleavage of a β-propiolactone with a dialkylaluminum acetylide to construct the α-sidechain. The synthesis began by conversion of the Corey lactone into enyne 1096. Treatment of 1096 with six equivalents of both

CHO

OSi

ÖTHP

1025

(EtO)$_2$P(O)CCl$_2$Li

Cl

Cl

OSi

ÖTHP

2 n-BuLi

H

OSi

ÖTHP

1096

n-butyllithium and dimethylaluminum chloride and nine equivalents of β-prop-iolactone at —40 → —30 °C followed by esterification and deprotection gave ester 1097, which was easily transformed into the title carbacyclin 1098.

G Ryanodine and the Ryanoid Insecticides

The ground stemwood of the plant *Ryania speciosa* has been used as an insecticide for over thirty years. An active constituent was isolated and identified to be the alkaloid ryanodine 1099 [496], the ester of pyrrole-α-carboxylic acid and ryanodol (1100). The complex structure of 1099 was elucidated by chemical degradation [497] and later confirmed by X-ray crystallography [498]. It was also shown that ryanodol (1100) could easily be converted to anhydroryanodol (1101). The

Deslongchamps group began synthetic studies towards ryanodine several years ago [499], culminating in the total synthesis of ryanodol (1100) in 1979 [500].

Their initial aim was the synthesis of the much simpler target 1101 where they hoped to construct rings B, C, and D via the strategy indicated in Scheme XCIII.

1102

1104   1103

Ozonolytic cleavage of the double bond in 1102 followed by intramolecular aldol condensation would lead to 1103, which could be crafted into 1104 with the correct choice of groups X, Y, and Z. In addition, these extra functional groups X, Y, and Z would be necessary to complete the synthesis of rings A and D of anhydroryanodol (1101). Lactone 1105 was chosen as a suitable diene and its synthesis from 5,6-dimethoxyindane is summarized in Scheme XCIV.

1105

The optically active dienophile 1106 for the initial Diels-Alder reaction was derived from (+)-carvone (Scheme XCV). Diene 1105 was refluxed in benzene with a slight excess of the dienophile 1106 to give a 1:1 mixture of adducts 1107 a,b and 1108 a,b in quantitative yield. Base-catalyzed intramolecular aldol condensation gave a mixture of four diastereoisomers 1109 and 1110 (exo and endo) in which the endo epimers predominated (3:1). The crude mixture was deprotected under acidic conditions and subjected to basic hydrolysis in order to effect a

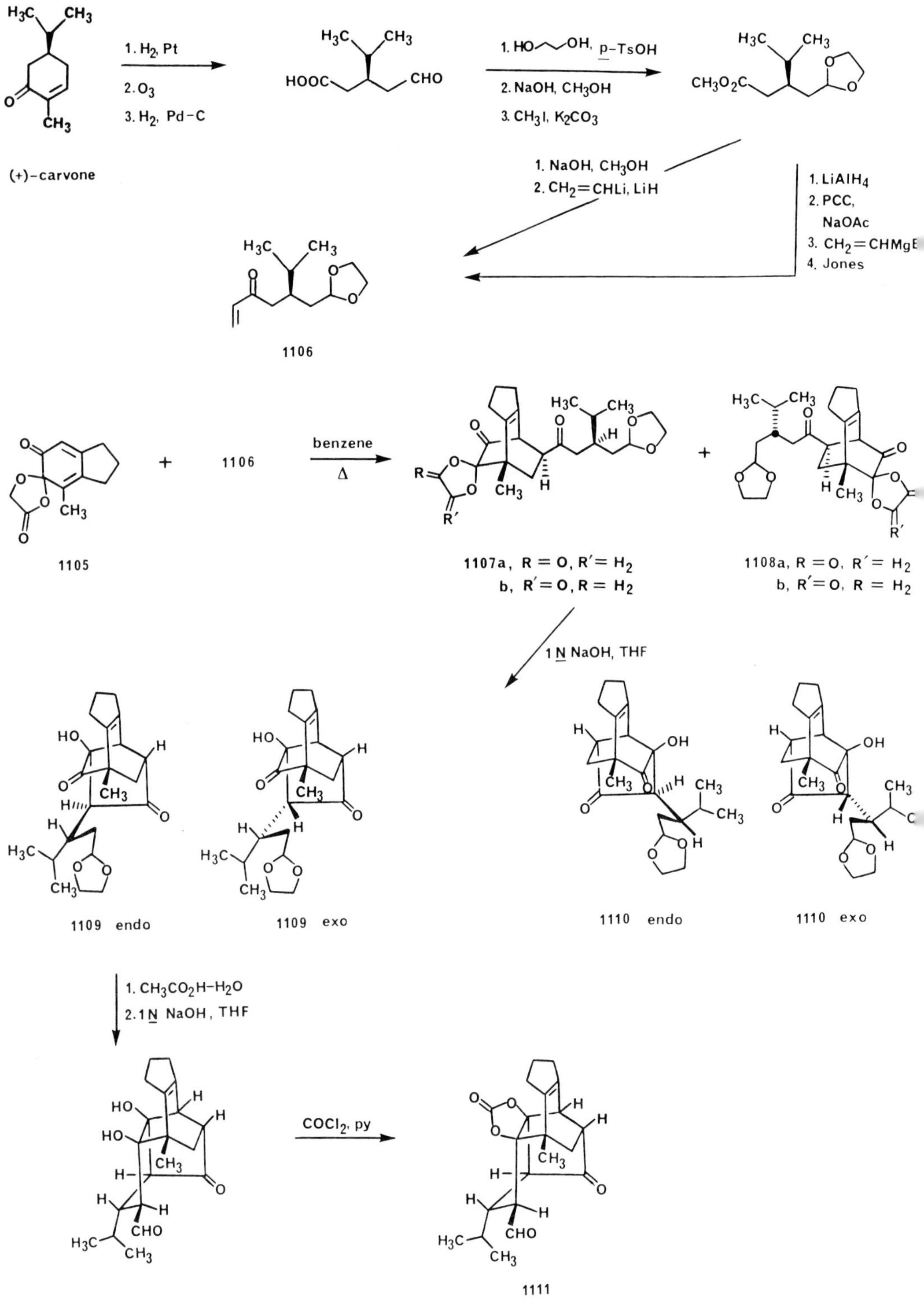

(+)-carvone
1. H_2, Pt
2. O_3
3. H_2, Pd–C
1. HO⌒⌒OH, p–TsOH
2. NaOH, CH_3OH
3. CH_3I, K_2CO_3
1. NaOH, CH_3OH
2. $CH_2=CHLi$, LiH
1. $LiAlH_4$
2. PCC, NaOAc
3. $CH_2=CHMgB$
4. Jones
1106
1105
benzene
Δ
1107a, R = O, R′ = H_2
b, R′ = O, R = H_2
1108a, R = O, R′ = H_2
b, R′ = O, R = H_2
1 N NaOH, THF
1109 endo
1109 exo
1110 endo
1110 exo
1. CH_3CO_2H–H_2O
2. 1 N NaOH, THF
$COCl_2$, py
1111

second intramolecular aldol condensation. After protection of the diol unit, the cyclic carbonate 1111 was isolated in 27% yield from diene 1105 (Scheme XCVI).

The next step in the strategy involved introduction of the lactone moiety followed by retroepoxidation to yield 1112, thereby setting the stage for the key ozonolysis-aldol condensation sequence. Introduction of the required methyl group at C-8 in 1113 and reduction of its ketone carbonyl to the equatorial alcohol were both achieved without difficulty. After cleavage of the lactone ring, triol 1114 underwent a highly selective oxidation to afford the hemiketal 1115. Mesylation followed by fragmentation under basic conditions furnished the large ring lactone 1116. Ring A of compound 1117 was further elaborated by conversion to a mixture of enol ethers followed by oxidation to the cyclopentanone product 1118. Alkaline hydrolysis of the endo allylic alcohol 1119 gave a separable mixture of anhydroryanodol (1101) and epianhydroryanodol (1120) (3:1). Epoxidation of either 1101 or 1120 gave the β-epoxide 1121 of epianhydroryanodol, which after reductive cyclization yielded ryanodal (1100) (Scheme XCVII).

167

Waterhouse and coworkers have isolated and characterized a new alkaloid 9,21-didehydroryanodine (1122) [501] as a constituent of *Ryania speciosa*. Reduction of 1122 gave both ryanodine (1099) and 9-epiryanodine (1123) in a 1:9 ratio. Recently, the same group has undertaken a structural examination of ryanodine and 9,21-didehydroryanodine by fully coupled two-dimensional ^{1}H—^{13}C shift correlation nuclear magnetic resonance spectroscopy [502].

IX Synthesis of Triquinane Natural Products

A Linear Triquinanes

1 Hirsutene

Hirsutene (1124) is the simplest member of a group of fungal metabolites possessing the linearly fused *cis,anti,cis*-tricyclo[6.3.0.0^{2,6}]undecane carbon skeleton [503]. As a result of the antibiotic and antitumor properties shown by some of these triquinanes, synthetic chemists have strived to devise new annulation procedures for their construction. Several new syntheses of hirsutene itself have recently been reported where a wide range of different strategies to provide access to the tricyclopentanoid ring system are described.

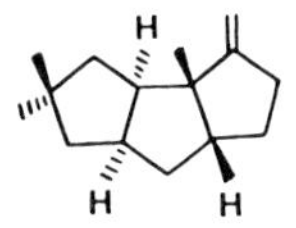

1124

The Stothers approach to (±)-hirsutene from dicyclopentadiene features homoketonization of the cyclopropoxide 1125 to the ring-expanded ketone 1126, followed by skeletal rearrangement under homoenolization conditions to provide

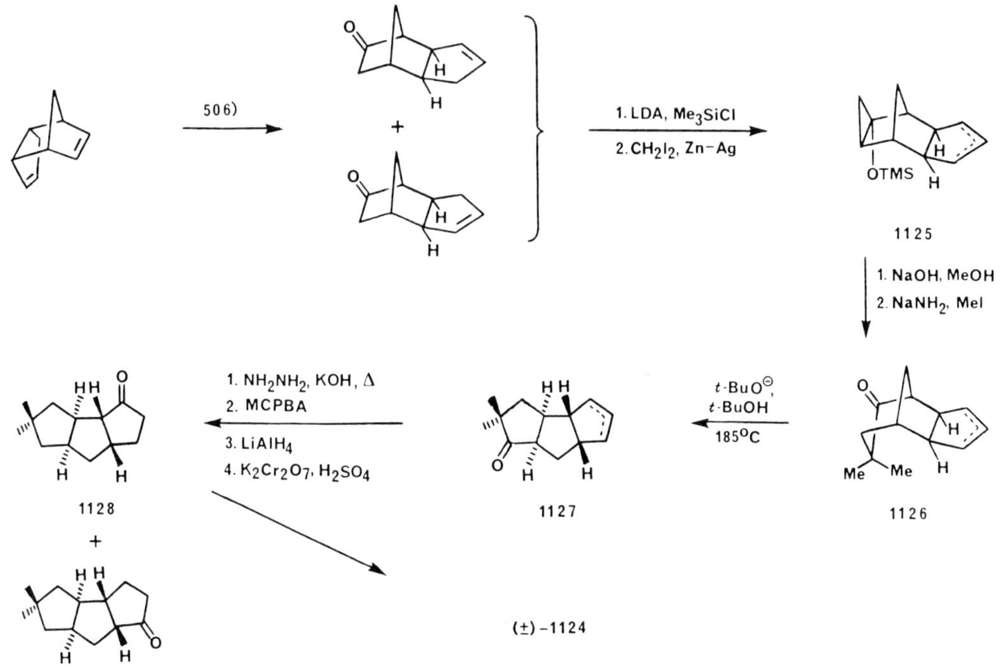

(±) -1124

169

the bicyclo[3.3.0]octan-2-one 1127 [504]. Following conversion to ketone 1128, which had previously been converted to (±)-hirsutene [503, 505], the formal synthesis was complete (Scheme XCVIII).

The key step in a previously reported synthesis of (±)-hirsutene by Little and coworkers [505] involved construction of triquinane 1130 via an intramolecular 1,3-diyl trapping reaction of the activated diylophile 1129. This approach required subsequent removal of the carbomethoxy activating group. In an attempt to improve their route, the same reaction has been carried out using the unactivated diylophile 1131 to produce a mixture of tricyclopentanoids 1132 and 1133 in a ratio of 5:1. Olefin 1132 was readily converted into ketone 1128 using a standard hydroboration-oxidation sequence, thereby completing a much shorter formal synthesis of hirsutene (Scheme XCIX) [507].

Funk and coworkers chose to employ an intramolecular nitrone-olefin cyc-loaddition in their stereospecific construction of the linearly-fused cyclopen-tanoid framework of hirsutene [508, 509]. Heating the isomeric mixture of N-methyl nitrones 1135/1136, obtained from ketone 1134, gave only the cycload-duct 1137. After methylation and hydrogenation, the resulting amino alcohol 1138 underwent reductive deamination via a Cope elimination to afford the olefin 1139. Oxidation and hydrogenation afforded known ketone 1128 (Scheme C).

Full details of a previously reported [510] hirsutene synthesis by the Magnus group have been published [511]. Thus, the bicyclic enone precursor 1142 was obtained via a new annulation strategy that utilizes reaction of acid chloride 1140 with vinyl silane 1141 in the presence of silver borofluoride.

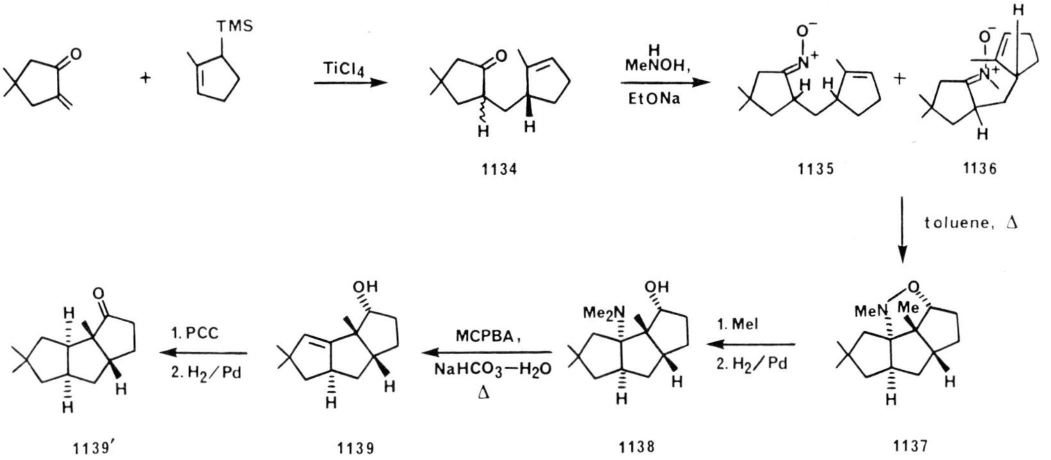
TMS
$TiCl_4$
H
MeNOH,
EtONa
1134
1135
1136
toluene, Δ
1. PCC
2. H_2/Pd
MCPBA,
$NaHCO_3$–H_2O
Δ
1. MeI
2. H_2/Pd
Me_2N
MeN
Me
OH
1139′
1139
1138
1137

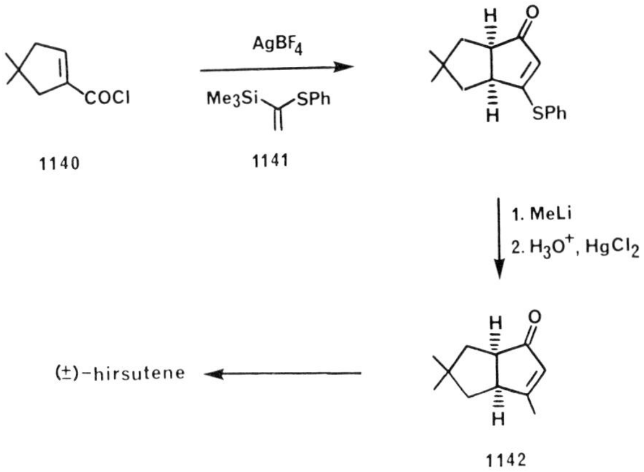
$AgBF_4$
COCl
Me_3Si
SPh
1140
1141
1. MeLi
2. H_3O^+, $HgCl_2$
(±)-hirsutene
1142

A short, efficient route to (±)-hirsutene developed by Curran employed tandem radical cyclization as the key step [512, 513]. Diiodide 1144, obtained in a number of steps from lactone 1143, was treated with lithium trimethylsilylacetylide, and subsequently desilylated to produce the cyclization precursor 1145. Treatment of 1145 with tri-*n*-butyltin hydride gave (±)-hirsutene (1124) in 83% yield after purification. The overall synthesis is summarized in Scheme CI.

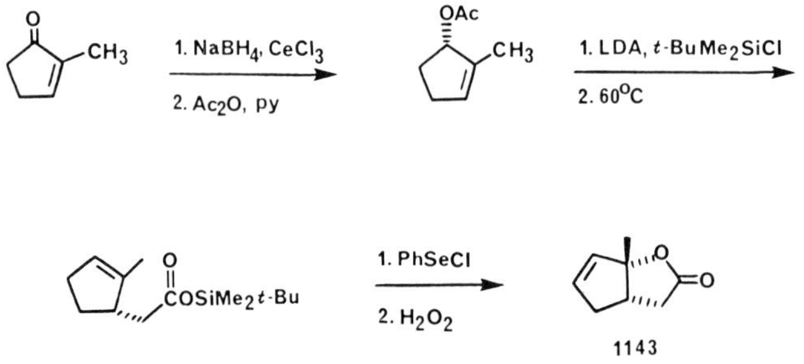
CH_3
1. $NaBH_4$, $CeCl_3$
2. Ac_2O, py
OAc
1. LDA, *t*-$BuMe_2SiCl$
2. 60°C
$COSiMe_2$*t*-Bu
1. PhSeCl
2. H_2O_2
1143

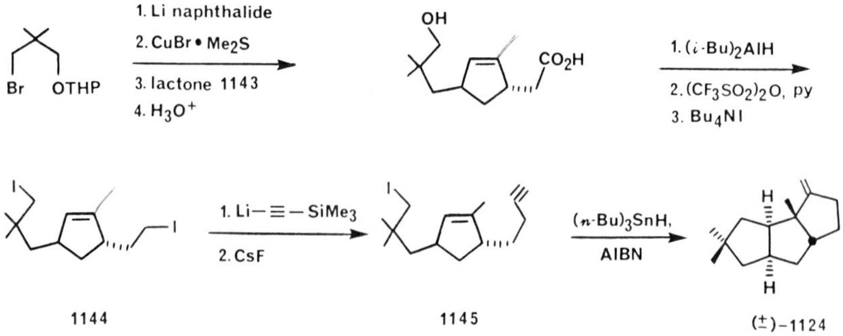

1144   1145   (±)-1124

In an asymmetric synthesis of (+)-hirsutene [514], Hua chose to utilize a condensation reaction between the anion derived from (+)-(R)-allyl p-tolyl sulfoxide and 2-methyl-2-cyclopentenone to afford after acetylation the 1,4-adduct 1146 with an optical purity of 96% ee. After reduction, intramolecular cyclization of the resulting enol acetate with the vinylic sulfide moiety gave a mixture of hexahydropentalenones 1147 and 1148 in a 4:1 ratio. The remaining ring as in 1128 was constructed by 1,4-addition of cuprate 1150 to enone 1149, followed by desilylation and tosylation. Upon treatment with sodium hydride in refluxing dimethoxyethane, 1151 underwent ring closure to triquinane 1152

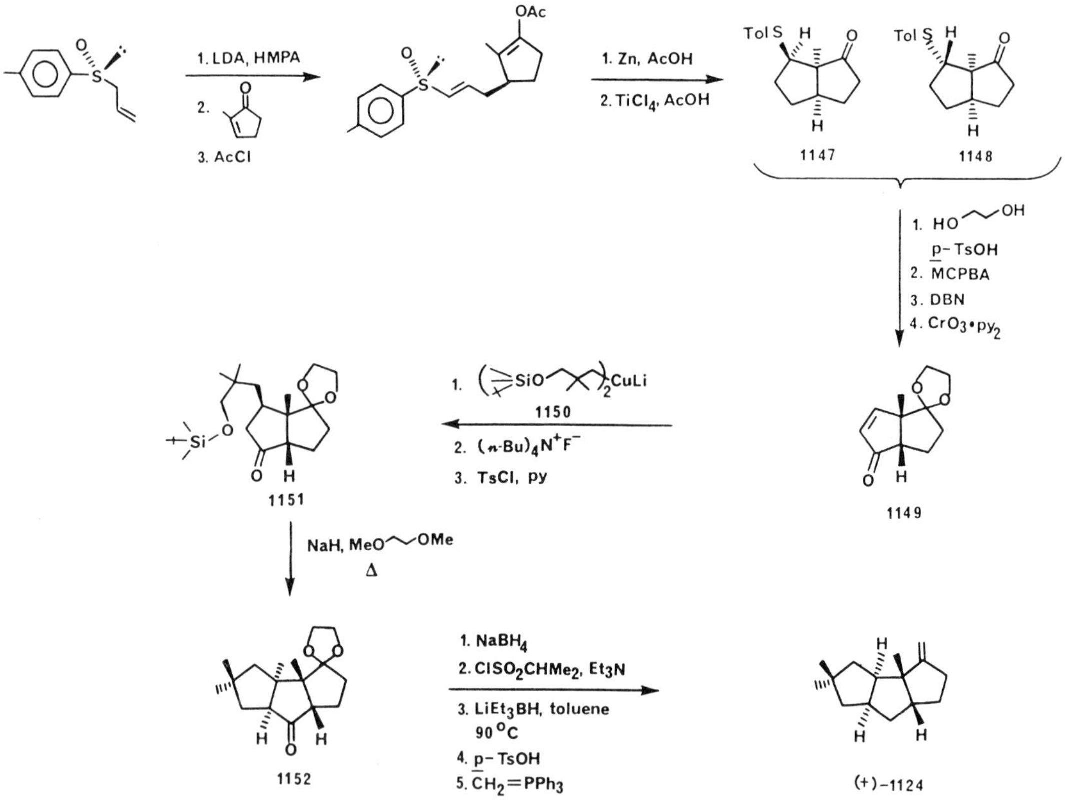

1147   1148

1151   1150   1149

1152   (+)-1124

(Scheme CII). Advancement to (+)-hirsutene proceeded without difficulty, the overall synthesis establishing the absolute configuration of (+)-1124 to be as shown.

The full details of the Ley approach to (±)-hirsutene [515], featuring an organoselenium-mediated cyclization reaction, have recently been reported [516].

In a short approach to (±)-hirsutene [517], Weedon and coworkers used the photochemical cycloaddition of the enol of 5,5-dimethylcyclohexane-1,3-dione to 2-methylcyclopent-2-enol followed by a low-valent titanium reduction [518] of the silylated adducts to afford 1153 and 1154 in 17% and 38% yields, respectively. Access to the ketone 1128 by sequential deprotection, hydrogenation, and oxidation completed their formal synthesis.

Hewson and coworkers gained access to the highly functionalized bicyclo[3.3.0]octane 1157 by means of a sequence involving intramolecular Wittig reaction of diketoester 1155 with vinyl phosphonium salt 1156 [519]. After dimethylation, demethoxycarbonylation, and Wolff-Kishner reduction, the vinyl sulfide thus obtained (1158) was oxidized to the corresponding sulfone. Deprotonation with n-butyllithium followed by quenching with iodo acetal 1159 gave 1160

173

1160

1. 6% Na–Hg,
   Na$_2$HPO$_4$
2. HCO$_2$H

1. 6% Na–Hg,
   Na$_2$HPO$_4$
2. pyridinium
   toluene-p–
   sulfonate

1161

RuCl$_3$•6 H$_2$O

(±)–1124

1139′

in high yield. Desulfenylation and concomitant double bond isomerization followed by deprotection led to aldehyde 1161 which has previously been converted to hirsutene. Scheme CIII illustrates the route as well as an alternative approach for construction of the C-ring.

2 The Capnellene Group

(−)-△$^{9(12)}$-Capnellene (1162), a marine sesquiterpene isolated in 1978 by Djerassi et al [520], is the parent hydrocarbon [521] and presumed biosynthetic precursor of the capnellanes, a group of polyhydroxylated derivatives [522–525] of 1162 (viz 1163 a—f) isolated from the same source, namely the soft coral Capnella

1162

1163a; R^1 = R^2 = R^3 = R^4 = H
  b; R^1 = R^2 = R^3 = H, R^4 = OH
  c; R^1 = R^2 = R^4 = H, R^3 = OH
  d; R^1 = R^3 = R^4 = H, R^2 = OH
  e; R^1 = R^3 = H, R^2 = R^4 = OH
  f; R^2 = R^4 = H, R^1 = R^3 = OH

imbricata. Although details of their biological profile are not known, it has been suggested that the capnellanes act as chemical defense agents to inhibit the growth of microorganisms [526, 527] and to prevent larval settlement [528]. Interest in their biological acitivity and structural novelty has caused the capnellane group to be the focal point of extensive synthetic efforts over the past few years.

By iterative application of an α-alkynone cyclization, Dreiding and coworkers have devised an interesting synthesis of the parent hydrocarbon 1162 [529]. α-Alkynone 1165 was prepared by acylation of bis(trimethylsilyl)acetylene with acid chloride 1164 under Lewis acid catalysis. On thermolysis, cyclization occurred to give bicyclic enone 1166. After elaboration of the more complex α-alkynone 1167, thermolysis furnished a 45:55 mixture of the linear and angular triquinanes 1168 and 1169, respectively. Since 1168 had previously been converted to $\triangle^{9(12)}$-capnellene [530, 531], the formal synthesis was complete (Scheme CIV).

Little and coworkers have presented a brief mechanistic rationale [532] for the unusual regiochemical result observed in the key intramolecular 1,3-diyl trapping reaction in their previously reported synthesis of $\triangle^{9(12)}$-capnellene [530].

Full details of the synthesis of epiprecapnelladiene (1174) reported by Pattenden and Birch [533, 534] have appeared. Thus, intramolecular photocycloaddition of the enol benzoate 1170 provided access to the tricyclic intermediate 1171. After dimethylation, the resulting keto benzoate 1172 underwent saponification and retroaldolization in ethanolic potassium hydroxide solution to give the cyclooctane-1,5-dione 1173. The transannular cyclization of epiprecapnelladiene 1174 had previously been reported by the same group (Scheme CV) [535].

In a novel approach to 1162, the Mehta group [536] thermolyzed pentacyclic dione 1175 to produce the triquinane bis-enone 1176 in 60% yield. Relocation of the double bond was effected by treatment with DBU to afford the new enone

1177. Standard reactions applied to 1177 led to the known nor-capnellenone 1178 [530, 531], thereby completing the short formal synthesis (Scheme CVI). A detailed account of the previously reported Paquette synthesis [531] of $\triangle^{9(12)}$-capnellene has recently been published [537].

The strategy employed in the Stille approach to 1162 involved an iterative three-carbon annulation procedure to construct two of the three fused rings [538].

Regiospecific palladium-catalyzed carbonylative coupling of enol triflates 1179 and 1182 with (trimethylsilyl)vinylstannane afforded the pair of key intermediates 1180 and 1183. Subsequent silicon-directed Nazarov cyclization [539] occurred under Lewis acid catalysis to provide the bicyclic and tricyclic systems 1181 and 1184, respectively. The remaining steps for conversion of 1184 to 1162 have previously been reported. The synthesis is summarized in Scheme CVII.

Starting from 2-cyclopentenone, Liu and Kulkarni initiated a synthesis of $\Delta^{9(12)}$-capnellene featuring a combination of [2 + 2] and [4 + 2] cycloaddition reactions (Scheme CVIII) [540]. Expansion of the cyclobutanone ring in 1185 was achieved upon treatment with ethyl diazoacetate in the presence of boron trifluoride etherate. After dehydrogenation of major product 1186 to enone 1188, Lewis acid catalyzed Diels-Alder cycloaddition to isoprene gave tricyclic ester 1189 in 60% yield. Functional group manipulation led to 1190, which underwent facile ring contraction when subjected to ozonolysis and reductive workup. Enone 1190 was converted to the corresponding oxime and subjected to Beck-

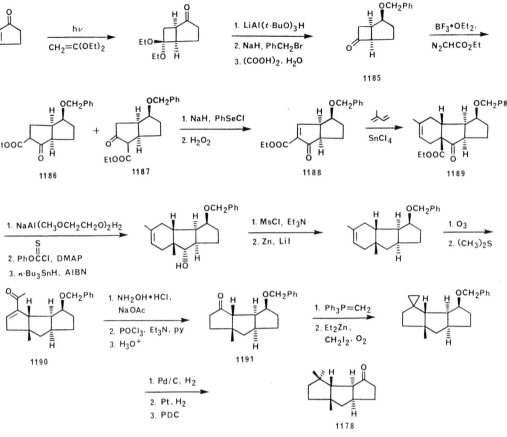

mann rearrangement with phosphorus oxychloride and subsequent acidification. The resulting ketone 1191 was easily transformed into the known intermediate 1178.

Using an efficient tandem radical cyclization strategy, Curran has more recently also realized a total synthesis of (±)-$\triangle^{9(12)}$-capnellene [541]. Bromide 1193, obtained in several steps from lactone 1192, was treated with tri-n-butyltin hydride in refluxing benzene containing AIBN to afford 1162 in 61% yield.

In an application of their recently reported methylenecyclopentane annulation sequence [542], the Piers group has reported an interesting synthesis of (±)-$\triangle^{9(12)}$-capnellene [543] (Scheme CIX). The cuprate reagent derived from 4-chloro-2-(trimethylstannyl)-1-butene was made to undergo conjugate addition to 2-methylcyclopentenone in order to obtain the chloro ketone 1194. Cyclization was effected by treatment with potassium hydride, leading to the bicyclic ketone 1195 in 75% yield. After futher elaboration to enone 1196, the identical annulation procedure was reemployed to construct the tricyclic ketone 1197, a convenient precursor of 1162.

The key step in the Grubbs approach [544] to (±)-$\triangle^{9(12)}$-capnellene involved rearrangement to the cyclobutene enol ether 1201 via the metallocycle 1199. The four asymmetric centers of capnellene were initially set up by means of a single intramolecular cycloaddition of the functionalized cyclopentadiene to produce 1198. Reaction with Tebbe's reagent [545] and 4-dimethylaminopyridine gave metallocycle 1199, which underwent cycloreversion to 1200 upon heating. Subsequent intramolecular trapping of 1200 resulted in the complete conversion of 1198 to 1201. Ring expansion of 1202, employing the usual conditions of ethyl diazoacetate catalyzed by boron trifluoride, provided the correct tricyclic skeleton of capnellene (Scheme CX).

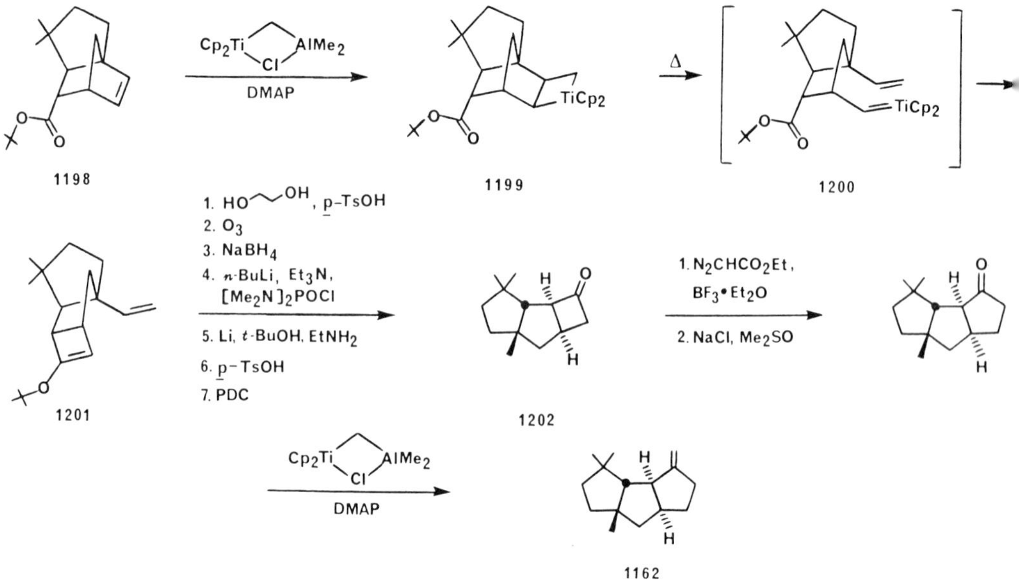

Cp_2Ti Cl $AlMe_2$
DMAP
$TiCp_2$
Δ
$TiCp_2$
1198
1199
1200
1. HO⌒OH, p-TsOH
2. O_3
3. $NaBH_4$
4. n-BuLi, Et_3N, $[Me_2N]_2POCl$
5. Li, t-BuOH, $EtNH_2$
6. p-TsOH
7. PDC
1. N_2CHCO_2Et, $BF_3 \cdot Et_2O$
2. NaCl, Me_2SO
1201
1202
Cp_2Ti Cl $AlMe_2$
DMAP
1162

In their synthetic studies directed toward the capnellane alcohols, the Ikegami group developed an efficient method for the construction of the bisallylic alcohol unit common to this group of compounds. Their strategy was applied to the

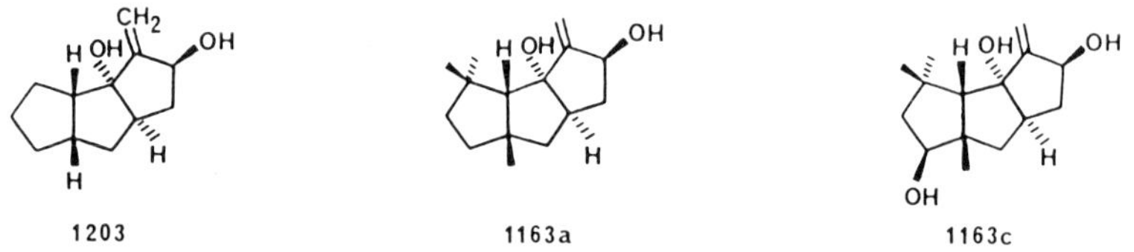

CH_2
H OH
OH
1203
1163a
1163c

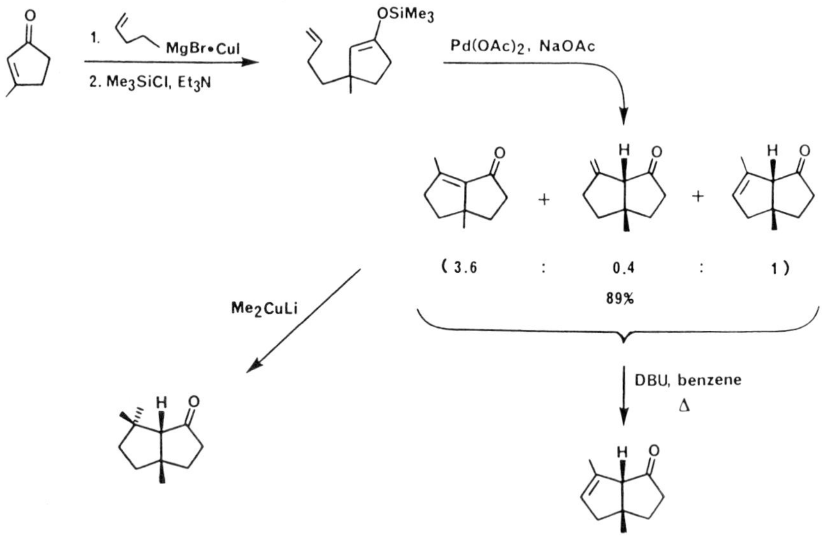

1. MgBr•CuI
2. Me_3SiCl, Et_3N
$OSiMe_3$
$Pd(OAc)_2$, NaOAc
(3.6 : 0.4 : 1)
89%
Me_2CuLi
DBU, benzene
Δ

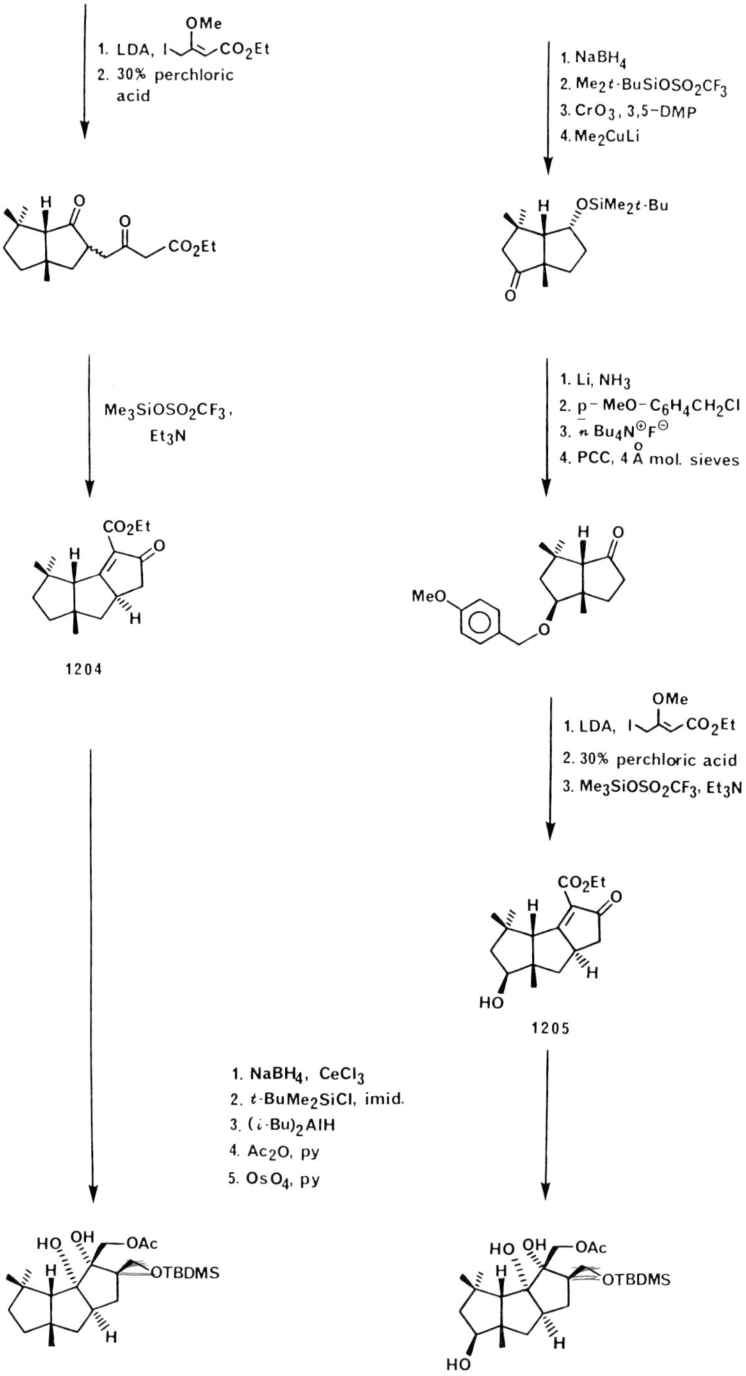
OMe
1. LDA, I–CH2–C(OMe)=CH–CO2Et
2. 30% perchloric acid
H O
O
CO2Et
Me3SiOSO2CF3, Et3N
CO2Et
H
O
H
1. NaBH4
2. Me2t-BuSiOSO2CF3
3. CrO3, 3,5-DMP
4. Me2CuLi
H
OSiMe2t-Bu
O
1. Li, NH3
2. p-MeO-C6H4CH2Cl
3. n Bu4N⊕F⊖
4. PCC, 4 Å mol. sieves
H O
MeO
O
OMe
1. LDA, I–CH2–C(OMe)=CH–CO2Et
2. 30% perchloric acid
3. Me3SiOSO2CF3, Et3N
CO2Et
H
O
H
HO
1. NaBH4, CeCl3
2. t-BuMe2SiCl, imid.
3. (i-Bu)2AlH
4. Ac2O, py
5. OsO4, py
HO OH
OAc
OTBDMS
H
H
HO OH
OAc
OTBDMS
H
H
HO

1204

1205

1. K_2CO_3
2. $MeSO_2Cl$, Et_3N
3. DBU
4. Me_3SiLi, HMPA
5. $(n\text{-}Bu)_4N^{\oplus}F^{\ominus}$

HO
OH

1163a

HO
OH

HO

1163c

synthesis of the model compound 1203 [546] and later to the first total synthesis of both $\triangle^{9(12)}$-capnellene-8β, 10α-diol (1163a) and $\triangle^{9(12)}$-capnellene-3β, 8β, 10α-triol (1163c, Scheme CXI) [547]. The routes to 1163a and 1163b are very similar, the latter employing an allylic oxidation with chromic anhydride and 3,5-dimethylpyrazole to introduce oxygen functionality at C-3. Intramolecular aldol cyclization in the presence of trimethylsilyl trifluoromethanesulfonate gave access to the tricyclic compounds 1204 and 1205, which were each converted through the same sequence of reactions to the target diol and triol.

3 Coriolin

Coriolin (1206), first isolated in 1969 from fermentation broths of the basidiomycete *Coriolus consors* [548], is a member of the hirsutane class of sesquiterpenes. The antibiotic and antitumor activities of coriolin and diketocoriolin B (1207), as well as their challenging structural features, continue to attract the attention of synthetic chemists.

HO
H
O
O
C B A
H
O
HO

1206

$OCO(CH_2)_6CH_3$
H
O
O
O
H
O

1207

In the Schuda synthesis of 1206 [549], the B and C rings were constructed from methanoindene 1208. Saturation of the double bond, methylation, and carbonyl reduction gave rise to an intermediate that ultimately provided 1209 (Scheme CXII). Annulation of the A-ring was carried out by Claisen alkylation, as developed by Lansbury [550], involving ketones 1210a and 1210b and vinyl chloride 1211 to afford 1212. Debenzoylation, hydrolysis of the vinyl chloride, and elimination of the propylthio group occurred simultaneously during treatment with mercuric acetate in 88% formic acid to deliver 1213. Aldolization and dehydration led to the known enone 1214, which has previously been converted to coriolin by several groups [551–553].

Magnus found occasion to utilize the Pauson-Khand reaction in an efficient formal synthesis of coriolin [554] (Scheme CXIII). Appropriate annulation of 1215 occurred upon treatment with dicobalt octacarbonyl in the presence of carbon monoxide and heptane at 110 °C in a sealed tube to afford a separable mixture of the bicyclo[3.3.0]octenone 1216a and its C-8 epimer 1216b (ca 3:1

ratio). Hydrogenation followed by alkylation with allyl bromide gave 1217, which upon Wacker oxidation provided 1,4-diketone 1218. Aldol cyclization of 1218 led to triquinane 1219. Upon arrival at enone 1220, the formal synthesis was complete since the last two steps have previously been described by Trost [554] and Danishefsky [552], respectively. In a short synthesis of (±)-coriolin [556], Koreeda and Mislanker utilized alkylation of dianion 1221 to construct the diquinane enone 1222 in 65% yield. Intermediate enone 1223 has previously been transformed into 1206 [553, 557] (Scheme CXIV).

184

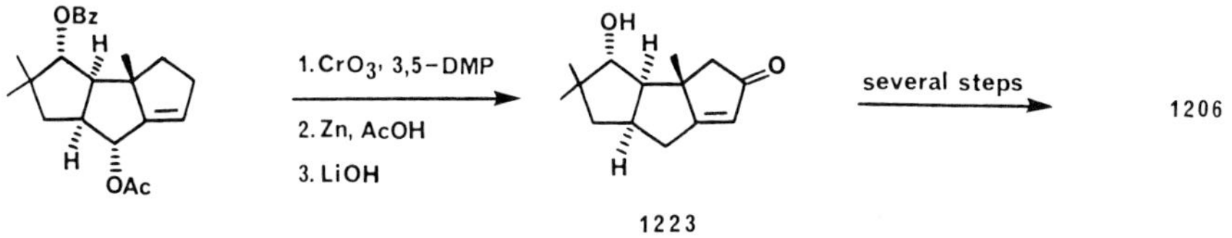

1223

The Wender approach to coriolin [558] featured a clever arene-olefin poly-cyclization strategy previously adopted in the synthesis of several polycyclic sesquiterpenes. In the first of the two routes, intramolecular photocyclization of 1-(2,6-dimethylphenyl)-1-acetoxy-2,2-dimethylpent-4-ene (1224) permitted stereospecific introduction of all of the groups pendant to two of the three target

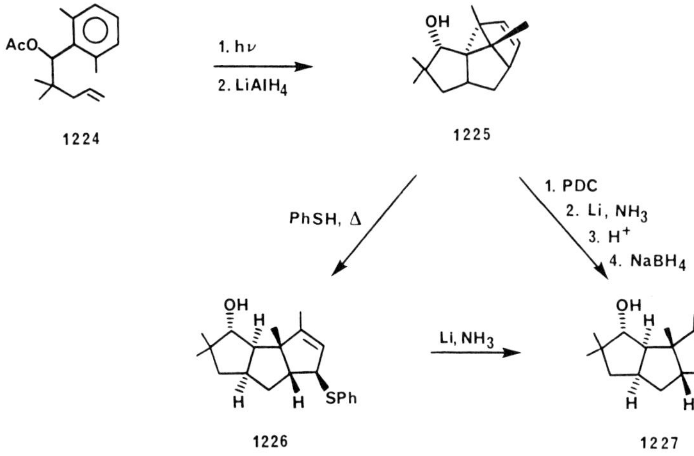

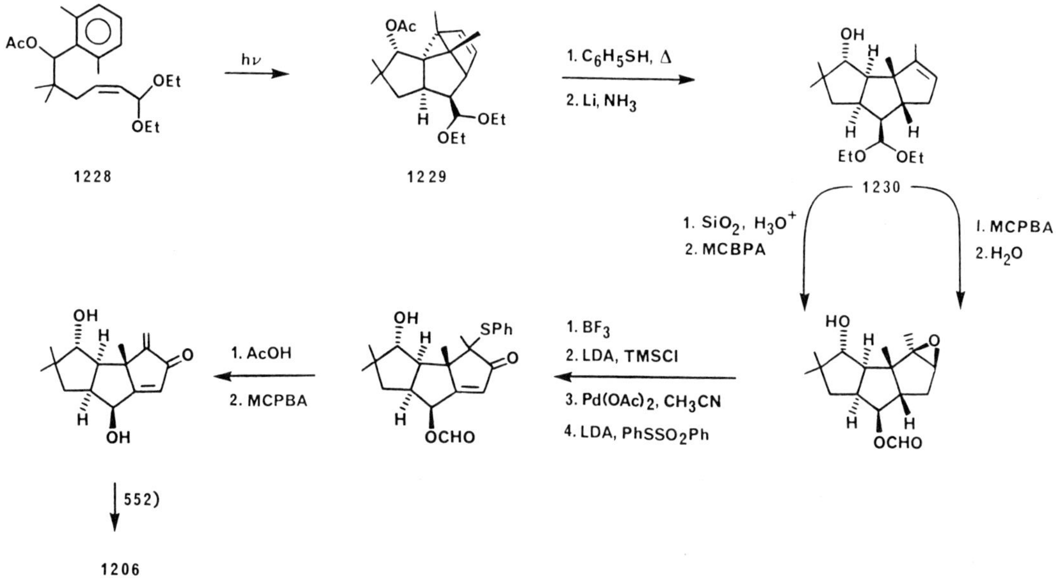

185

rings, providing 1225 as the major product after deacetylation. Heating 1225 with thiophenol induced a 1,5-radical addition to give 1226 as the sole product. Reductive desulphurization secured 1227, an intermediate also obtained in a four-step sequence starting from 1225. In the second route, 1228 was photocyclized to afford the tricyclic acetate 1228 (Scheme CXV). Treatment as before with thiophenol followed by reduction gave triquinane 1230. The overall synthesis of coriolin is summarized in Scheme CXV.

Full details of the Matsumoto approach [557] to coriolin from dicyclopentadiene have recently been reported [559].

The key step in the Demuth synthesis of coriolin [560] involved the regioselective photochemical rearrangement of β,γ-unsaturated ε-diketone epimers 1231 a, b (Scheme CXVI). After alkylation of the 1232 a, b mixture with methylallyl chloride, a single diketone (1223) was obtained. Reduction with lithium in liquid ammonia effected one-pot regiospecific cleavage of the cyclopropane ring and stereospecific C-6 carbonyl reduction. Osmium tetroxide-periodate oxidation of 1234 and aldol condensation led to the known enone 1235. The formal synthesis was completed by oxone oxidation to diol 1220, a well established coriolin precursor [552].

Employing a previously reported strategy illustrated in the synthesis of hirsutene, Funk and coworkers have also accomplished a formal synthesis of coriolin [509] via the same nitrone-olefin intramolecular cycloaddition of intermediates 1135 and 1136. Intermediate 1139, the synthesis of which is summarized

186

in Scheme C was crafted into coriolin by means of a short sequence of steps. Diene 1236 was obtained by pyrolysis of the thionocarbonate derived from alcohol 1139. Two subsequent hydroboration-oxidation reactions gave diols 1237 a, b and these were converted to the known enone 1235 [557, 559] (Scheme CXVII).

Little has once again employed an intramolecular 1,3-diyl trapping reaction, earlier utilized in his syntheses of hirsutene [507] and $\triangle^{9(12)}$-capnellene [532] to construct ($\pm$)-coriolin [561]. Following preparation of diazene 1238 from dihydro-5-(hydroxymethyl)-4,4-dimethyl-2(3H)-furanone, its facile conversion to the linearly fused tricyclopentanoid alcohol 1239 was effected under photolytic conditions. Acquisition of the known enone 1240 completed the formal synthesis of 1206 (Scheme CXVIII).

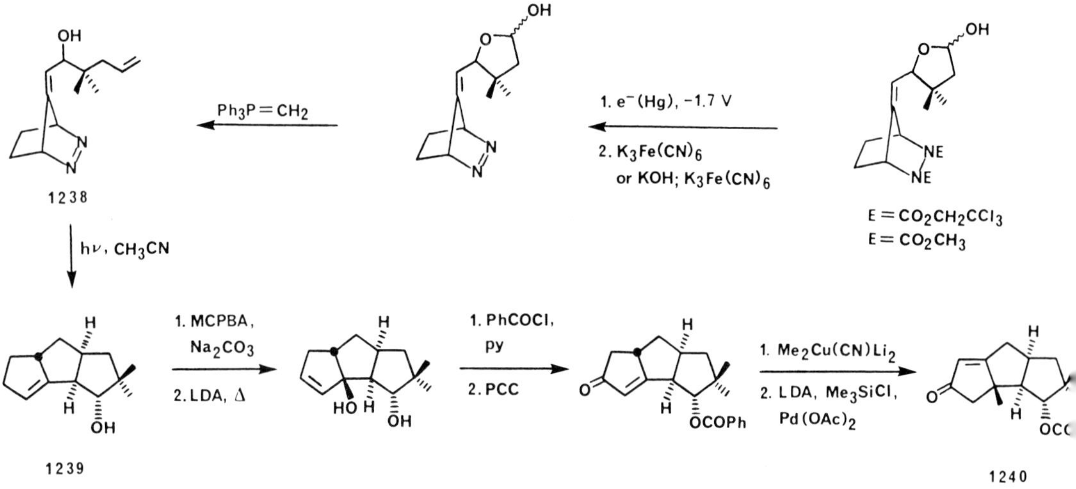

1238

1239

1240

4 Hirsutic Acid

Hirsutic acid C (1243), isolated from *Stereum hirsutum* [562] and *Stereum complicatum* [563], is a member of the hirsutane group of sesquiterpenes. Like the other members of this group, which includes the coriolins and hirsutene, 1243 has been a popular synthetic target.

Ikegami and coworkers, who previously achieved syntheses of racemic [564] and (+)-hirsutic acid [565], have more recently reported a modification to their asymmetric route [566]. The alternative, stereoselective route to intermediate

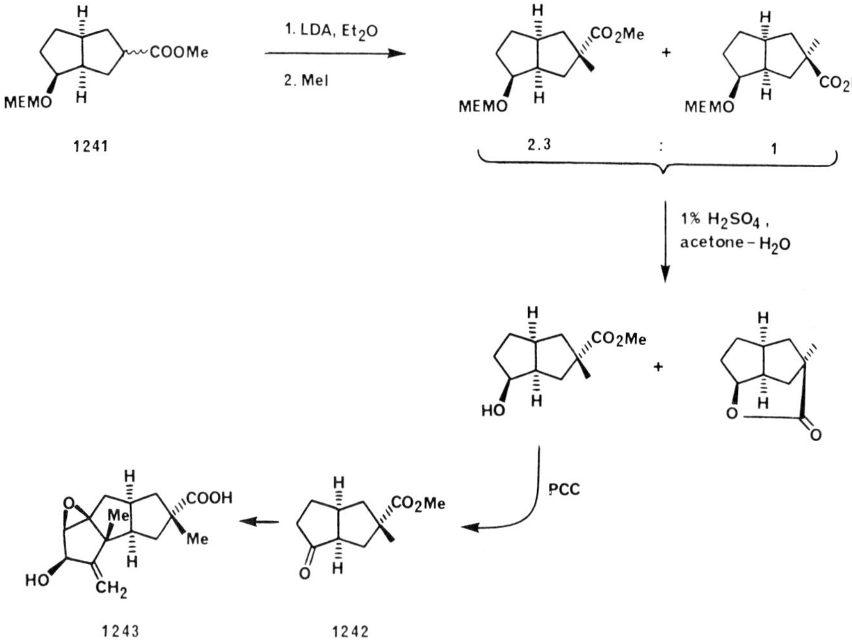

1241

2.3 : 1

1242

1243

1242, previously converted to (+)-hirsutic acid, featured chelation-controlled methylation from the sterically more crowded face of ester 1241 (Scheme CXIX).

The Greene synthesis of (±)-1243 [567] employed an iterative olefin annulation strategy to construct the second and third rings sequentially. Dichloroketene cycloaddition to cyclopentene 1244 afforded a separable mixture of 1245 and 1246 in a 3:1 ratio. Keto acid 1245 was treated with lithium dimethylcopper to generate the α-chloro enolate, which in the presence of excess methyl iodide and hexamethylphosphoric triamide provided the corresponding α-chloro α-methyl ketone. Ring expansion and esterification were effected with diazomethane. After reduction and elimination of 1247, the resulting olefin (1248) underwent [2 + 2] dichloroketene cycloaddition as before. Repetitive diazomethane ring enlargement and reduction led to triquinane 1249 (Scheme CXX).

189

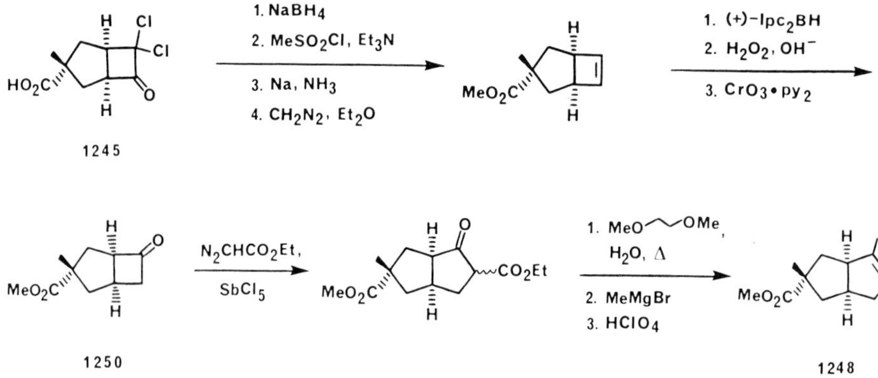

1245

1250

1248

In a synthesis of (+)-hirsutic acid [568], the same group employed an asymmetric hydroboration followed by oxidation to obtain the chiral ketone (−)-1250 in 90% ee. After advancement to bicyclo[3.3.0]ketone 1248, the synthesis linked up with the earlier racemic route [567].

Employing a strategy earlier utilized in their synthesis of coriolin, Magnus and coworkers have developed a route to (±)-hirsutic acid [554] where the key step involved a Pauson-Khand cyclization of 1251 to obtain 1252 and 1253 in a 55:45 ratio (Scheme CXXI). After desilylation, equilibration, hydrogenation, and

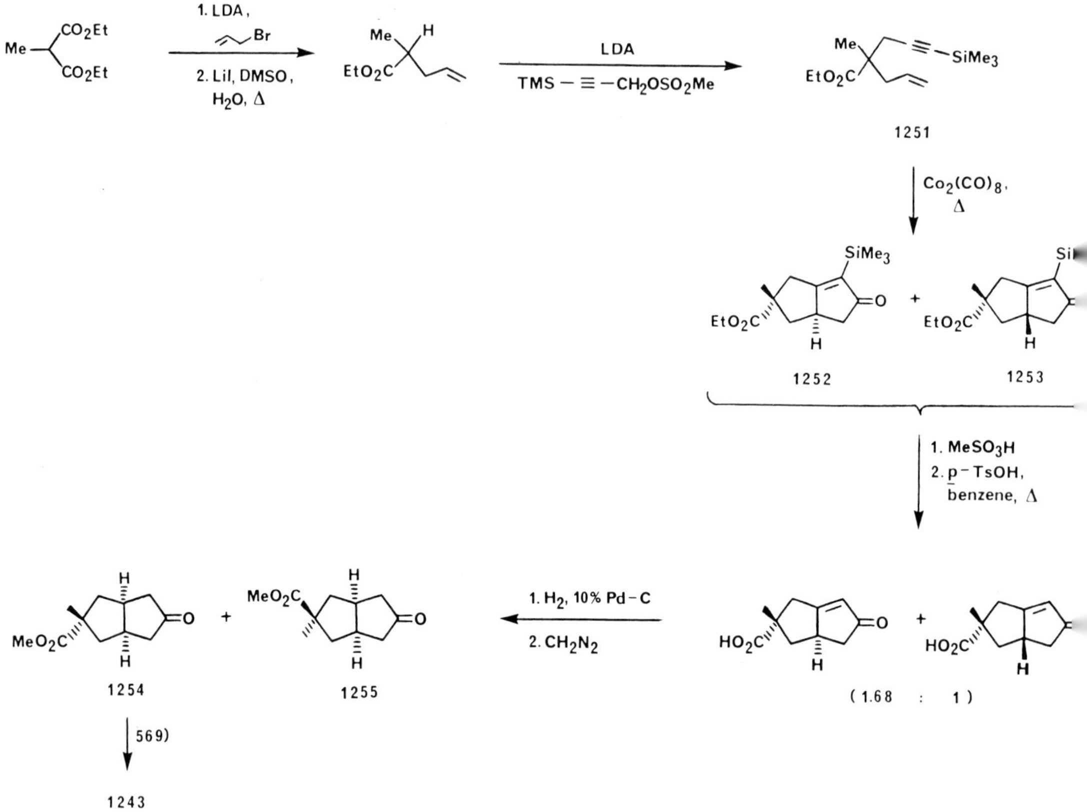

1251

1252   1253

1. MeSO$_3$H
2. p−TsOH, benzene, Δ

1254   1255

(1.68 : 1)

569)

1243

esterification, isomers 1254 and 1255 (1.7:1) were obtained. Since 1254 was also the Matsumoto precursor to hirsutic acid [569], the formal synthesis was complete.

B Angular Triquinanes

1 Isocomene Sesquiterpenes

Isocomene (1256) is a sesquiterpene hydrocarbon belonging to the group of angular triquinanes [570–572]. Wenkert and coworkers have now reported the full details of their route to 1256 which deploys a crucial α-oxycyclopropylcarbinol to cyclobutanone rearrangement [573].

1256

In 1983, Hudlicky reported the stereocontrolled synthesis of (±)-isocomenic acid (1266) and (±)-epiisocomenic acid (1256) involving as the key step a vinylcyclopropane to cyclopentene rearrangement (Scheme CXXII) [574]. Initially, the (1:1) *E:Z* mixture of acids 1257 was converted to the corresponding diazoketones 1258 and heated with a slurry of copper sulphate and copper(II) acetylacetonate in benzene. Flash vacuum pyrolysis of 1259 gave the angular triquinane 1260. After reduction, the resulting ketone was homologated by Wittig reaction with methylenetriphenylphosphorane at room temperature to afford a mixture of 1261 and 1262. If the reaction temperature was raised, 1261

underwent isomerization to 1262. The esters 1263 and 1264 were converted to the corresponding acids 1265 and 1266, respectively.

In a separate report, Hudlicky accomplished the conversion of esters 1261 and 1262 to isocomene (1256), epiisocomene (1269), β-isocomene (1268) and β-epiisocomene (1267) [575], the details of which are summarized in Scheme CXXIII.

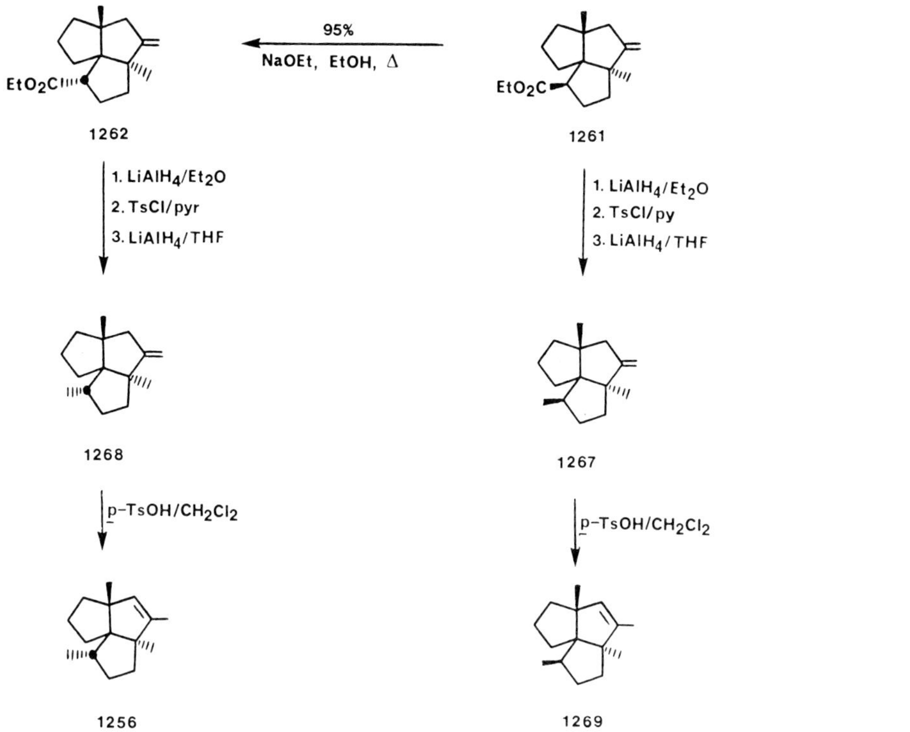
95%
NaOEt, EtOH, Δ
EtO2C
EtO2C
1262
1261
1. LiAlH4/Et2O
2. TsCl/pyr
3. LiAlH4/THF
1. LiAlH4/Et2O
2. TsCl/py
3. LiAlH4/THF
1268
1267
p-TsOH/CH2Cl2
p-TsOH/CH2Cl2
1256
1269

With the modified acid 1270 as starting material, the same group achieved an alternative synthesis of β-epiisocomene [576], which indicated that the original Chatterjee synthesis may have given 1269 and not 1256 as claimed [577].

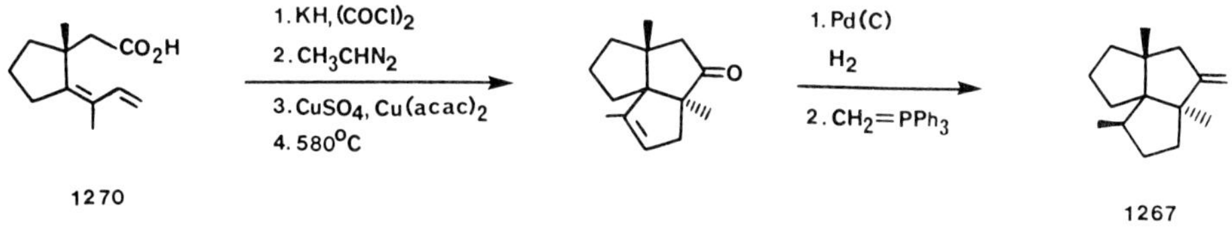
CO2H
1. KH, (COCl)2
2. CH3CHN2
3. CuSO4, Cu(acac)2
4. 580°C
O
1. Pd(C)
H2
2. CH2=PPh3
1270
1267

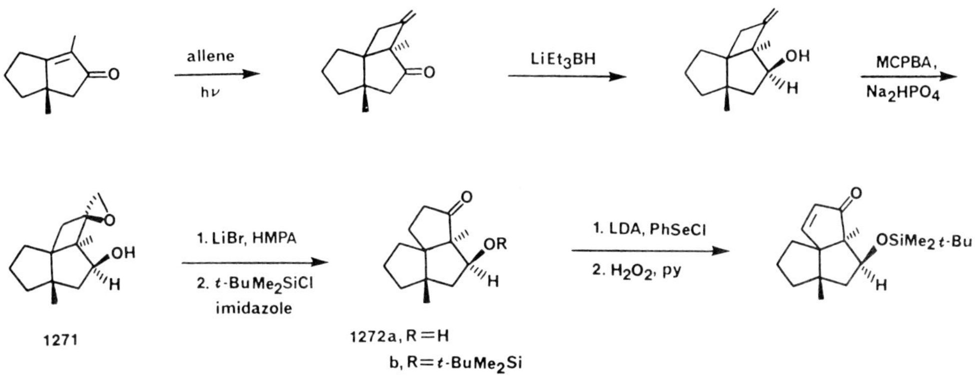
O
allene
hν
O
LiEt3BH
OH
H
MCPBA,
Na2HPO4
O
OH
H
1. LiBr, HMPA
2. t-BuMe2SiCl
imidazole
O
OR
H
1. LDA, PhSeCl
2. H2O2, py
O
OSiMe2t-Bu
H
1271
1272a, R=H
b, R=t-BuMe2Si

1. LiMe₂Cu → ... (scheme)

1. LiMe₂Cu
2. N₂H₄·H₂O, K₂CO₃

1. PCC
2. Ph₃MePBr, EtC(Me)₂ONa

1268

1256

The Tobe synthesis of isocomene (1256) and β-isocomene (1268) employed a chelation-controlled regioselective epoxide-carbonyl rearrangement as the key step [578] (Scheme CXXIV). Thus, treatment of epoxide 1271 lithium bromide and hexamethylphosphoric triamide in benzene at 80 °C proceeded regioselectively to furnish hydroxy ketone 1272 a possessing the correct carbon skeleton for both natural products.

2 Silphinene

Silphinene, the first member of a new family of triquinane natural products, was initially reported by Bohlmann and Jakupovic in 1980 [579]. The first two successful approaches to silphinene by Paquette [580] and by Ito [581] have been followed more recently by two additional syntheses.

In a ten-step route to (±)-silphinene (1275), Sternbach and coworkers chose to employ an intramolecular Diels-Alder reaction of the substituted cyclopentadiene 1273 [582] to produce tricyclic olefin 1274 (Scheme CXXV). Compound 1274 was elaborated into the natural product in four steps.

Na⁺Cp⁻

Li(OEt)C=CH₂

1273

160 °C
benzene,
sealed
tube

(10 : 1)

H⁺

1274

1. O₃
2. Me₂S

1. KOH
2. Jones

Pd(OAc)₄, py
Cu(OAc)₂

(7 : 3)

1. Me₂CuLi
2. N₂H₄•H₂O, K₂CO₃

1275

Wender and Ternansky have developed an impressive three-step synthesis of 1275 employing arene-olefin meta photocycloaddition strategy [583]. Irradiation of 1276, obtained from commercially available starting materials, with Vycor-filtered light from a mercury arc lamp afforded a 1:1 mixture of 1277 and 1278 in 40% yield. Reductive cleavage of the cyclopropane ring in 1277 with lithium in monomethylamine at −78 °C gave (±)-silphinene (1275) along with its positional isomer 1279 in a ratio of 9:1.

1276   1277   1278

Li, MeNH₂

1275   1279

3 Pentalenene and Pentalenic Acid

(+)-Pentalenene (1280), isolated from *Streptomyces griseochromogenes* in 1980 [584], is the parent hydrocarbon of the pentalenolactone family of antibiotic fungal metabolites. Synthetic interest has arisen due to its novel biosynthetic origin [585] and its role as a key precursor pf pentalenolactone (788). In recent years, several new syntheses of 1280 have followed the first two reports by Matsumoto [586] and by Paquette [587].

1280   788

In the Piers approach to 1280 [588], a new annulation strategy effected the transformation of diquinane enone 1281 into the tricyclic ketone 1282 (Scheme CXXVI). After hydrogenation, the resulting mixture of ketones 1283a and 1283b were treated with methyllithium followed by acid-catalyzed elimination to afford a separable mixture of ($\pm$)-pentalenene (1280) and its C-9 isomer 1284.

1281

1283a, $R^1 = Me$, $R^2 = H$
   b, $R^1 = H$, $R^2 = Me$

1282

1280. $R^1 = Me$, $R^2 = H$
1284, $R^1 = H$, $R^2 = Me$

Pattenden and Teague chose to employ a strategy based on transannular cyclization of bicyclo[6.3.0]undecadiene 1287 in the presence of boron trifluoride etherate (Scheme CXXVII) [589]. Conversion of dione 1285 to its silyl enol ether and [2 + 2] photocycloaddition afforded the tricyclic ketone 1286. After cuprate addition, the resulting tertiary alcohol was converted into a ring-expanded bicyclic ketone by treatment with hydrofluoric acid in aqueous tetrahydrofuran.

1285   1286

1280   1287

The overall strategy is very similar to the previously reported Pattenden synthesis of $\triangle^{9(12)}$-capnellene.

In their synthesis of the pentalenene precursor 1289, Baker and Keen utilized a palladium(O)-catalyzed [3 + 2] cycloaddition to construct the diquinane ketone 1288 [590]. Precursor 1289 could easily be converted to 1290 obtained earlier in the Paquette synthesis of 1280 [587] (Scheme CXXVIII).

1288

1289   1290

Starting from 1,5-dimethylcycloocta-1,5-diene, Mehta developed a short route to pentalenene [591] that features a transannular cyclization as the key step, somewhat similar to the Pattenden approach [589]. Base-catalyzed epimerization of 1291 gave a 4:1 mixture of cis- and trans-isomers 1292 and 1291; the former upon aldol condensation led to bicyclic enone 1293. Transannular closure produced triquinane ketone 1294 in 55% yield. These and the final steps are outlined in Scheme CXXIX.

Pentalenic acid (1295) [592–4] has also been identified as a key intermediate in the biosynthesis of pentalenolactone. It is thought that the antitumor compound deoxypentalenic acid glucuron (1296a) [595] may also be biosynthetically related to 788, 1280, and 1295, although this has not been proven.

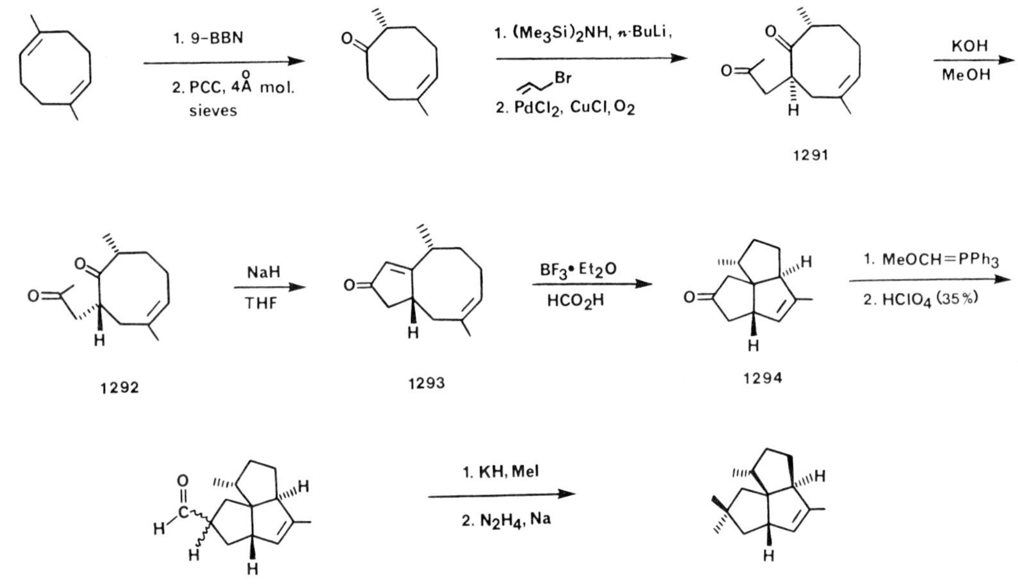

1291

1292   1293   1294

1280

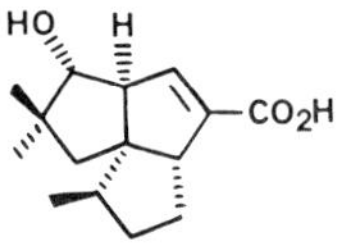

1295 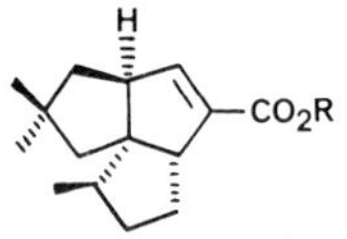   1296a, R =

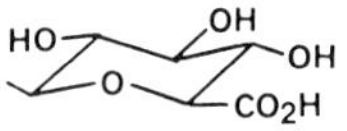

b, R = H

Crimmons and DeLoach have achieved syntheses of 1280, [596] 1295 [596, 597] and 1296 b [597] via a strategy that features an intramolecular photocycloaddition-cyclobutane fragmentation. To this end, photolysis of diester 1297 gave a mixture of tricyclic products 1298 a, 1298 b, and 1299 in a 10:3:1 ratio. Cleavage of the cyclobutane ring in 1298 a, b with lithium in liquid ammonia afforded a separable 13:1 mixture of spiro-fused esters 1300 a and 1300 b. After conversion of 1300 a into tricyclic keto alcohol 1301, the three targets were easily obtained as summarized in Scheme CXXX.

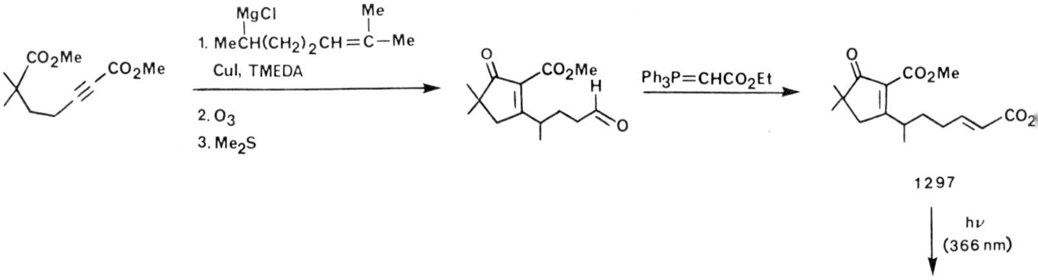

1297

hν
(366 nm)

1. HCl, H$_2$O, AcOH, Δ

2. MeOH, p–TsOH, (MeO)$_3$CH

CO$_2$Me

HO CO$_2$Me CO$_2$Et

R$_1$
R$_2$

1300a, R$_1$=Me, R$_2$=H
   b, R$_1$=H, R$_2$=Me

Li, NH$_3$

O CO$_2$Me CO$_2$Et

1298a,b

O CO$_2$Me CO$_2$Et

+

1299

t-BuO$^{\ominus}$K$^{\oplus}$

1. p–TsOH, HO⌒OH

2. Li, NH$_3$

3. 10% HCl

OH H O

1301

1. LDA, CO$_2$

2. HCl

3. CH$_2$N$_2$

OH H O CO$_2$Me

1. NaBH$_4$
2. Ac$_2$O, Et$_3$N, DMAP
3. DBU
4. KOH

1. MeSO$_2$Cl, Et$_3$N
2. DBU

OH CO$_2$Me

1. LDA, CO$_2$
2. H$^+$
3. CH$_2$N$_2$
4. NaBH$_4$

O

1295

1. H$_2$, Pd–C
2 MeSO$_2$Cl, Et$_3$N
3. DBU
4. KOH

1296b

1. LDA, MeI
2. Li, NH$_3$

OH

1. p–tol–O–C(=S)–Cl

2. 200°C, 200mm Hg

1280

4 Senoxydene

In 1979, Bohlmann and Zdero isolated a new sesquiterpene known as senoxydene from *Senecio oxyodontus* [598]. On the basis of spectral characteristics, it was assigned the structure 1302. A few years later, the Paquette group reported an unambiguous, stereocontrolled synthesis of 1302 which proved that senoxydene did not possess this structure [599]. An identical conclusion was later arrived at independently by Ito and coworkers [600]. The synthetic 1302 differed in its ^{1}H

1302

NMR spectrum from the natural product; in particular, the four methyl signals deviated substantially from the reported values. In an attempt to discover if this was due to an alternative location of the gem-dimethyl group, Paquette and coworkers synthesized isomers 1306 and 1310 [601], but neither of these compounds proved identical to natural senoxydene. The routes to these isomeric hydrocarbons illustrated in Scheme CXXXI follow a strategy similar to that adopted earlier for the preparation of 1302 [599]. More specifically, the final five-membered ring was introduced via alkylation of ketones 1303 and 1307 with (E)-1-iodo-2-(trimethylsilyl)but-2-ene to afford adducts 1304 and 1308, respectively. After transformation into the corresponding 1,4-diketones 1305 and 1309, cyclization occurred under basic conditions. Conventional functional group manipulation ultimately led to the target compounds.

1303

1304

1305

1306

1307

1308

1309

1310

Ito and coworkers began their synthesis of alleged "senoxydene" [600] with the key intermediate 1311 used earlier in their route to silphinene [581]. Tri-

CrO₃, HOAc–H₂O

1. HO⌒OH, H⁺
2. LiOH

1. MeLi
2. ⌒MgCl
3. NaH, MeI, HMPA

1311

HOOC

HOOC

1. O₃
2. NaBH₄
3. H⁺, H₂O
4. MeSO₂Cl, py

t-BuOK, THF

MeLi

OMe

OMe   OMs

1312

OMe

1313

1. MeSO₂Cl, Et₃N
2. H₂, PtO₂, AcOH
3. BF₃·OEt₂

OMe

1314

1302

BF₃·OEt₂, p-TsOH

1. MCPBA
2. BF₃·OEt₂

1. LiAlH₄
2. MeSO₂Cl, Et₃N
3. LiAlH₄

1315

quinane ketone 1313 was obtained in 93% yield by intramolecular alkylation of 1312 (Scheme CXXXII). After the discovery that their synthetic senoxydene was different from the natural product, the same group carried out a synthesis of 1315 via 1314 where the secondary methyl group is in the β instead of the α configuration. However, this material was also not identical with senoxydene "1302".

The Asakawa group, in their hydroxylation studies of natural products using meta-chloroperbenzoic acid, examined the reaction of cedrol with this reagent [602]. One of the products (1316a) obtained in low yield was converted to the C-2

MCPBA
CHCl₃, 6h
Δ

+

1316a, R¹ =OH, R²=R³=R⁴=R⁵=H
b, R²=OH, R¹=R³=R⁴=R⁵=H
c, R³=OH, R¹=R²=R⁴=R⁵=H
d, R⁴=OH, R¹=R²=R³=R⁵=H
e, R⁵=OH, R¹=R²=R³=R⁴=H

1317

Pb(OAc)₄
(on a)

epimer 1318 of "alleged senoxydene" and the route is illustrated in Scheme CXXXIII. Their synthetic material proved identical to that obtained previously by Paquette and by Ito.

The "senoxydene problem" has still not yet been resolved, but it is clear that the structure proposed is incorrect and requires revision.

5 Silphiperfolene and Congeners

Silphiperfol-6-ene (1319) was isolated by Bohlmann and Jakupovic in 1980 from the roots of *Silphium perfoliatum* [603]. Shortly thereafter, the related ketone 5-oxosilphiperfol-6-ene (1320) was characterized as a constituent of the stems of *Espeletiopsis quacharaca* [604].

In 1984, the Paquette group reported the first syntheses of these interesting tricyclic sesquiterpenes [605]. Their approach, which began from (R)-(+)-pulegone [606] (Scheme CXXXIV), allowed assignment of absolute stereochemistry as shown in the formulas. Alkylation of keto ester 1321 with 1-(tosyloxy)-2-ethyl-2-propene followed by ozonolysis, aldol cyclization, and decarboxylation led to 1322. Marfat-Helquist annulation [607, 608] of 1322 gave aldol 1323, which was dehydrated and deoxygenated to produce the angular triquinane 1324. From this point, the elaboration of (−)-1319 proceeded without

difficulty. Since the pyridinium chlorochromate oxidation of (−)-1319 to give (−)-1320 had previously been accomplished [604], this work also comprised a formal total synthesis of (−)-5-oxosilphiperfol-6-ene.

Wender and Singh have described the total synthesis of (±)-silphiperfol-6-ene (1319), (±)-7βH-silphiperfol-5-ene (1329), and (±)-7αH-silphiperfol-5-ene (1330) by utilizing an arene-olefin metaphotocycloaddition [609] as the key step. The initial photolysis of 1325 led to a mixture of 1326 and its vinyl cyclopropane isomer 1327 in a 1:1.88 ratio. These two compounds were photochemically interconvertible. Thus, irradiation of 1326 in acetaldehyde provided a 60% yield of 1328 which was converted to 1319, 1329, and 1330 (Scheme CXXXV).

203

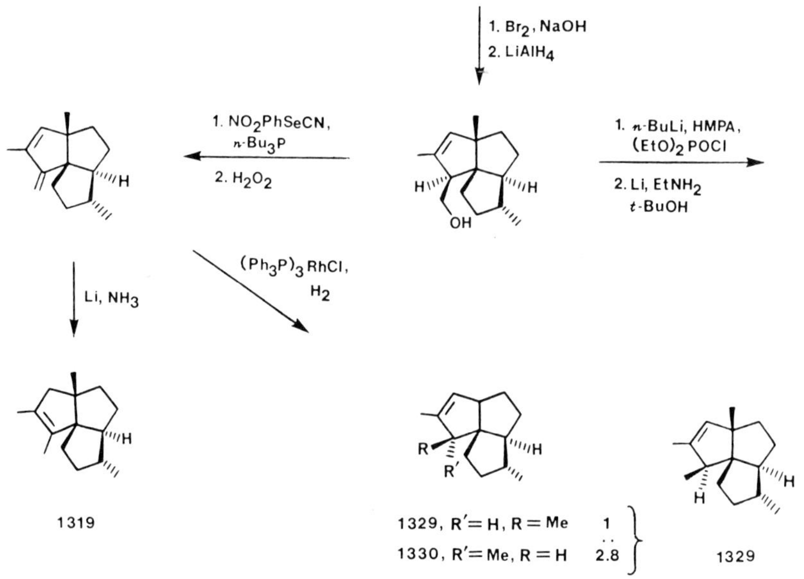

1319

1329, R′=H, R = Me   1
1330, R′=Me, R = H   2.8

1329

Curran and Kuo chose to utilize tandem radical cyclization in a short synthesis of (±)-silphiperfol-6-ene (1319) and (±)-9-episilphiperfol-6-ene (1336) [610]. The cyclization precursor of 1331 was readily available by sequential alkylation of 3-ethoxy-2-cyclopenten-1-one with methyl iodide and (E)-2-methyl-1,3-di-

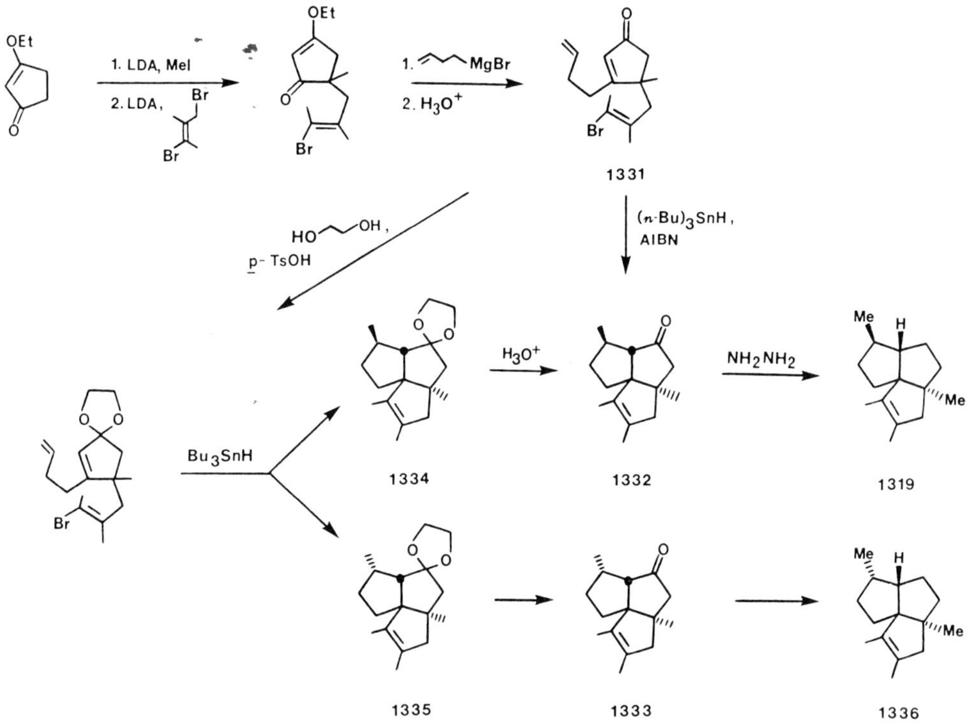

1331

1334   1332   1319

1335   1333   1336

bromo-2-butene, addition of butenylmagnesium bromide, and hydrolysis. Tin hydride-promoted ring closure furnished a difficultly separable (1:3) mixture of 1332 and 1333. However, it was found that prior protection of the carbonyl group gave rise to a more favorable ratio of 1334 and 1335 (2.5:1), which were now readily separable. Deprotection and Wolff-Kishner reduction provided the targeted molecules (Scheme CXXXVI).

6 Retigeranic Acid

The first total synthesis of (±)-retigeranic acid (1341), a novel sesterterpene monocarboxylic acid isolated from various lichens in the Himalayas [611, 612] was achieved recently by the Corey group [613]. This impressive synthesis

features several interesting reactions including the stereospecific construction of bicyclic ketone 1337, the Diels-Alder reaction of diene 1338 with methyl-3-formyl-*cis*-crotonate, a [2 + 2] ketene olefin cycloaddition to produce cyclobutanone adduct 1339, and ring expansion strategy to generate ketone 1340. The overall route summarized in Scheme CXXXVII is a brilliant example of the ongoing development of stereospecific chemical reactions in the construction of complex organic molecules.

C Propellane Structures

1 Modhephene

The novel sesquiterpene modhephene (1345) is the first carbocyclic [3.3.3]propellane to be isolated from natural sources [614] and thus it has been the target of several innovative syntheses in recent years [615–624].

In the Tobe approach [621] the key chelation-controlled regioselective rearrangement of epoxide 1342 gave rise to propellane 1343. It is noteworthy that migration of the less-substituted C-10 carbon of 1342 predominated, a possible

$$h\nu, \text{ allene}$$

(4.5 : 1)

1. Me₃SiOCH₂CH₂OSiMe₃, Me₃SiOSO₂CF₃
2. MCPBA, Na₂HPO₄
3. separation

LiBr, HMPA
C₆H₆, 80°C

1343   1342

1. MeMgI
2. SOCl₂, py

1. MCPBA, Na₂HPO₄
2. BF₃•OEt₂

1. N₂H₄•H₂O, K₂CO₃
2. H₂SO₄

1344

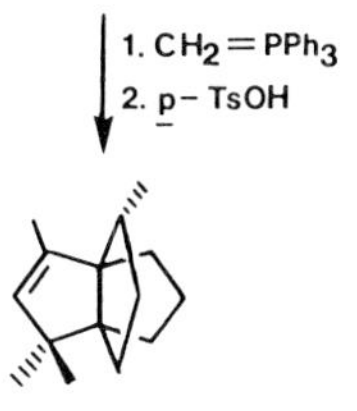

1. $CH_2 = PPh_3$
2. p-TsOH

1345

result of chelation between the lithium cation and acetyl oxygen atom. Triquinane 1343 was transformed in turn into ketone 1344, a recognized precursor of modhephene (1345) (Scheme CXXXVIII) [615, 616, 618].

Mundy and Wilkening achieved access to the Smith intermediate 1347 [615] in a formal synthesis of 1345 (Scheme CXXXIX) that features a dianion-mediated cyclopentannulation reaction followed by dehydration and subsequent

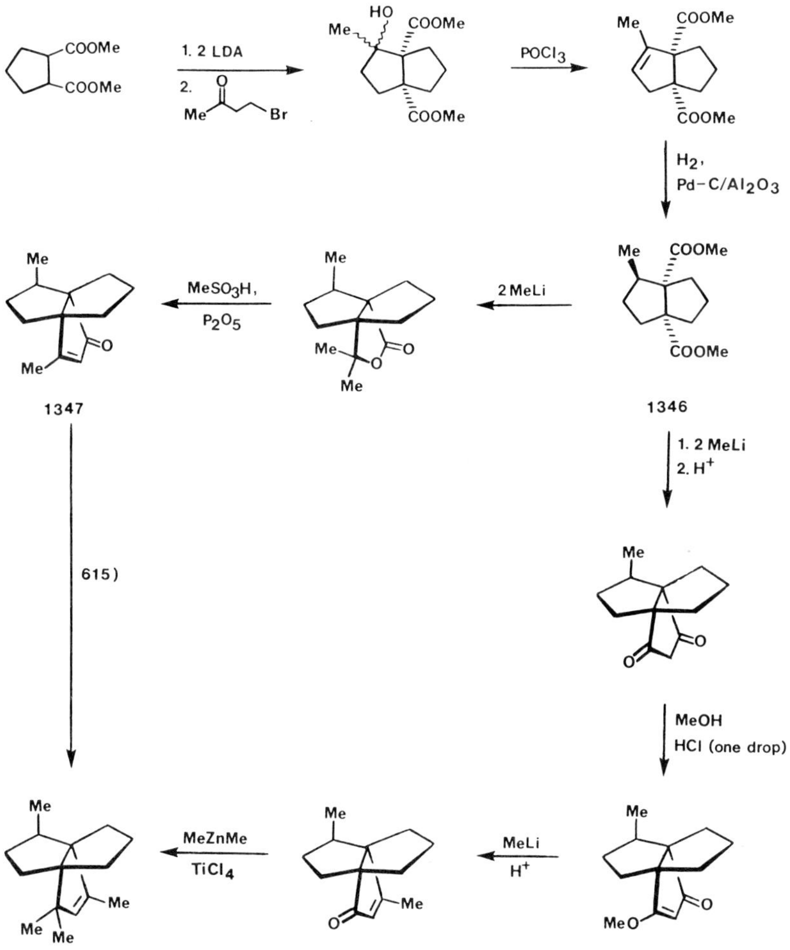

COOMe / COOMe → 1. 2 LDA / 2. Me—CO—CH₂—Br → HO, Me, COOMe, COOMe → $POCl_3$ → Me, COOMe, COOMe

H_2, Pd–C/Al_2O_3

Me, COOMe, COOMe

1346

2 MeLi ←

Me, Me, —O, O (lactone)

MeSO₃H, P_2O_5 ←

1347

615)

1. 2 MeLi
2. H^+

Me (diketone structure)

MeOH
HCl (one drop)

Me, MeO, O

MeLi / H^+ ←

Me, O, Me

MeZnMe / $TiCl_4$ ←

Me, Me, Me, Me

1345

207

heteroatom-assisted stereospecific hydrogenation to construct the diester 1346 [622]. In a more recent report [623], the details of this synthetic work, in addition to a new total synthesis of modhephene were disclosed. Thus, diester 1346 was converted via a four-step sequence into 1345 (Scheme CXXXIX).

A new approach to the [3.3.3]propellane system by Mehta and Subrahmanyam [624] featured photochemical oxa-di-π-methane rearrangement of tricyclic system 1348 to the strained tetracycle 1349. After regioselective reductive cleavage of 1350 and oxidation to afford propellane 1351, standard reactions led to the natural product (1345) in addition to its isomer 1352 (Scheme CXL).

X References

1. Paquette, L. A.: Topics in Current Chemistry 79, 41 (1979)
2. Paquette, L. A.: ibid. 119, 1 (1984)
3. Ramaiah, M.: Synthesis 529 (1984)
4. Trost, B. M.: Chem. Soc. Rev. 11, 141 (1982)
5. Lee, T. V., Richardson, K. A.: Tetrahedron Lett. 3629 (1985)
6. Trost, B. M., Fray, M. J.: ibid. 4605 (1984)
7. Dorsch, M., Jäger, V., Spönlein, W.: Angew. Chem., Int. Ed. Engl. 23, 798 (1984)
8. Aristoff, P. A.: Synth. Commun. 13, 145 (1983)
9. Geetha, G., Raju, N., Rajagopalan, K., Swaminathan, S.: Ind. J. Chem. 238 (1984)
10. Gaviña, F., Costero, A. M., Luis, S. V.: J. Org. Chem. 49, 4613 (1984)
11. Koreeda, M., Lian, Y., Akagi, H.: J. Chem. Soc., Chem. Commun. 449 (1979)
12. Wilkening, D., Mundy, B. P.: Synth. Commun. 14, 227 (1984)
13. Furuta, K., Ikeda, N., Yamamoto, H.: Tetrahedron Lett. 675 (1984)
14. Camps, P., Castane, J., Santos, M. T.: Chem. Lett. 1367 (1984)
15. Vollhardt, J., Gais, H.-J., Lukas, K. L.: Angew. Chem., Int. Ed. Engl. 24, 610 (1985)
16. Trost, B. M., Cossy, J., Burks, J.: J. Am. Chem. Soc. 105, 1052 (1983)
17. Miyashita, M., Yanami, T., Kumazawa, T., Yoshikoshi, A.: ibid. 106, 2149 (1984)
18. Nakashita, Y., Watanabe, T., Benkert, E., Lorenzi-Riatsch, L., Hesse, M.: Helv. Chim. Acta 67, 1204 (1984)
19. Duthaler, R., Maienfisch, P.: ibid. 67, 856 (1984)
20. Duthaler, R., Maienfisch, P.: ibid. 67, 845 (1984)
21. Hewson, A. T., MacPherson, D. T.: Tetrahedron Lett. 5807 (1983)
22. Bestmann, H. J., Schade, G., Lütke, H., Mönius, T.: Chem. Ber. 118, 2640 (1985)
23. Marino, J. P., Laborde, E.: J. Am. Chem. Soc. 107, 734 (1985)
24. Ishibashi, H., Okada, M., Komatsu, H., Ikeda, M., Tamura, Y.: Synthesis 643 (1985)
25. Crandall, J. K., Magaha, H. S.: J. Org. Chem. 47, 5368 (1982)
26. Kakiuchi, K., Takeuchi, H., Tobe, Y., Odaira, Y.: Bull. Chem. Soc. Jpn. 58, 1613 (1985)
27. Nokami, J., Wakabayashi, S., Okawara, R.: Chem. Lett. 869 (1984)
28. Corey, E. J., Pyne, S. G.: Tetrahedron Lett. 2821 (1983)
29. Pattenden, G., Robertson, G. M.: ibid. 4617 (1983); Tetrahedron 41, 4001 (1985)
30. Fox, D. P., Little, R. D., Baizer, M. M.: J. Org. Chem. 50, 2202 (1985)
31. Burks, J. E., Jr., Crandall, J. K.: ibid. 49, 4663 (1984)
32. Hafner, K., Thiele, G. F.: Tetrahedron Lett. 1445 (1984)
33. Corey, E. J., Kang, M.: J. Am. Chem. Soc. 106, 5384 (1984)
34. Beckwith, A. L. J., Schiesser, C. H.: Tetrahedron 41, 3925 (1985)

35. Beckwith, A. L. J., Roberts, D. H., Schiesser, C. H., Wallner, A.: Tetrahedron Lett. 3349 (1985)
36. Winkler, J. D., Sridar, V. J.: J. Am. Chem. Soc. *108*, 1708 (1986)
37. Leonard, W. R., Livinghouse, T.: Tetrahedron Lett. 6431 (1985)
38. Clive, D. L. J., Beaulieu, P. L., Set, L.: J. Org. Chem. *49*, 1313 (1984)
39. Angoh, A. G., Clive, D. L. J.: J. Chem. Soc., Chem. Commun. 980 (1985)
40. Belotti, D., Cossy, J., Pete, J. P., Portella, C.: Tetrahedron Lett. 4591 (1985)
41. Trost, B. M., Chan, D. M. T.: J. Am. Chem. Soc. *105*, 2315, 2326 (1983)
42. Trost, B. M., Nanninga, T. N., Satoh, T.: ibid. *107*, 721 (1985)
43. Shimizu, I., Ohashi, Y., Tsuji, J.: Tetrahedron Lett. 5183 (1984)
44. Danheiser, R. L., Carini, D. J., Fink, D. M., Basak, A.: Tetrahedron *39*, 935 (1983)
45. Saito, I., Shimozono, K., Matsuura, T.: Tetrahedron Lett. 5439 (1984)
46. Ashkenazi, P., Kettenring, J., Migdal, S., Gutman, A. L., Ginsburg, D.: Helv. Chim. Acta *68*, 2033 (1985)
47. Kubiak, G., Cook, J. M., Weiss, U.: J. Org. Chem. 49, 561 (1984)
48. Bertz, S. H., Kouba, J., Sharpless, N. E.: J. Am. Chem. Soc. *105*, 4116 (1983)
49. Kubiak, G., Cook, J. M., Weiss, U.: Tetrahedron Lett. 2163 (1985)
50. Sternbach, D. D.; Hughes, J. W., Burdi, D. F., Forstot, R. M.: ibid. 3295 (1983)
51. Alward, S. J., Fallis, A. G.: Can. J. Chem. *62*, 121 (1984)
52. Shishido, K., Hiroya, K., Fukumoto, K., Kametani, T.: Chem. Lett. 87 (1985)
53. Doyle, M. D., Trudell, M. L.: J. Org. Chem. *49*, 1196 (1984)
54. Callant, P., D'Haenens, L., Vandewalle, M.: Synth. Commun. *14*, 155 (1984)
55. Kreiser, W., Below, P.: Liebigs Ann. Chem. 203 (1985)
56. Schultz, A. G., Dittami, J. P., Eng, K. K.: Tetrahedron Lett. 1255 (1984)
57. Dauben, W. G., Rocco, V. P., Shapiro, G.: J. Org. Chem. *50*, 3155 (1985)
58. Wu, T.-C., Houk, K. N.: J. Am. Chem. Soc. *107*, 5308 (1985)
59. Christl, M., Brunn, E., Lanzendorfer, F.: ibid. *106*, 373 (1984)
60. Ferrier, R. J., Tyler, P. C., Gainsford, G. J.: J. Chem. Soc. Perkin I 295 (1985)
61. Mehta, G., Rao, K. S.: Tetrahedron Lett. 1839 (1984)
62. Sugihara, Y., Sugimura, T., Murata, I.: Bull. Chem. Soc. Jpn. *56*, 2859 (1983)
63. Greene, A. E. Charbonnier, F.: Tetrahedron Lett. 5525 (1985)
64. Paquette, L. A., Valpey, R. S., Annis, G. D.: J. Org. Chem. *49*, 1317 (1984)
65. (a) Knapp, S., Trope, A. F., Theodore, M. S., Hirata, N., Barchi, J. J.: J. Org. Chem. *49*, 608 (1984); (b) Hart, T. W., Comte, M.-T.: Tetrahedron Lett. 2713 (1985)
66. Chatani, N., Furukawa, H., Kato, T., Murai, S., Sonoda, N.: J. Am. Chem. Soc. *106*, 430 (1984)
67. Piers, E., Maxwell, A. R., Moss, N.: Can. J. Chem. *63*, 555 (1985)
68. Barbarella, G., Pitacco, G., Russo, C., Valentin, E.: Tetrahedron Lett. 1621 (1983)
69. Barbarella, G., Brückner, S., Pitacco, G., Valentin, E.: Tetrahedron *40*, 2441 (1984)
70. Tolstikov, G. A., Miftakhov, M. S., Danilova, N. A., Shitikova, O. V.: Zh. Org. Khim. *21*, 675 (1985)
71. Ogino, T., Kazama, T., Kobayashi, T.: Chem. Lett. 1863 (1985)
72. Ghosh, S., Saha, S.: Tetrahedron *41*, 349 (1985)
73. Pauson, P. L., Khand, I. U.: Ann. N. Y. Acad. Sci. *295*, 2 (1977)
74. Khand, I. U., Pauson, P. L., Habib, M. J. A.: J. Chem. Res. (S) 348, (M) 4418 (1978)
75. Daalman, L., Newton, R. F., Pauson, P. L., Wadsworth, A.: J. Chem. Res. (S) 346, (M) 3150 (1984)
76. Newton, R. F., Pauson, P. L., Taylor, R. G.: J. Chem. Res. (S) 277, (M) 3501 (1980)
77. Montana, A.-M., Moyano, A., Pericas, M. A., Serratosa, F.: Tetrahedron *41*, 5995 (1985)

78. Knudsen, M. J., Schore, N. E.: J. Org. Chem. *49*, 5025 (1984)
79. Smit, W. A., Gybin, A. S., Shashkov, A. S., Strychkov, Y. T., Kyz'mina, L. G., Milkaelian, G. S., Caple, R., Swanson, E. D.: Tetrahedron Lett. 1241 (1986)
80. Carceller, E., Centellas, V., Moyano, A., Pericas, M. A., Serratosa, F.: ibid. 2475 (1985)
81. Exon, C., Magnus, P.: J. Am. Chem. Soc. *105*, 2477 (1983)
82. Magnus, P., Principe, L. M.: Tetrahedron Lett. 4851 (1985)
83. Schultz, A. G., Dittami, J. P., Lavieri, F. P., Salowey, C., Sundararaman, P., Szymula, M. B.: J. Org. Chem. *49*, 4429 (1984)
84. Schultz, A. G., Lavieri, F. P., Snead, T. E.: J. Org. Chem. *50*, 3086 (1985)
85. Demuth, M., Mikhail, G.: Tetrahedron *39*, 991 (1983)
86. Demuth, M., Wietfeld, B., Pandey, B., Schaffner, K.: Angew. Chem., Int. Ed. Engl. *24*, 763 (1985)
87. Mehta, G., Subrahmanyam, D., Subba Rao, G. S. R., Pramod, K.: Ind. J. Chem. *24B*, 797 (1985)
88. Callant, P., DeWilde, H., Storme, P., Vandewalle, M.: Bull. Soc. Chim. Belg. *93*, 489 (1984)
89. Storme, P., Callant, P., Vandewalle, M.: ibid. *92*, 1019 (1983)
90. Demuth, M., Hinsken, W.: Angew. Chem., Int. Ed. Engl. *24*, 973 (1985)
91. Demuth, M., Cánovas, A., Weigt, E., Krüger, C., Tsay, Y.-H.: ibid. *22*, 721 (1983)
92. Drew, M. G. B., Gilbert, A., Heath, P., Mitchell, A. J., Rodwell, P. W.: J. Chem. Soc., Chem. Commun. 750 (1983)
93. Drew, M. G. B., Gilbert, A., Rodwell, P. W.: Tetrahedron Lett. 949 (1985)
94. Osselton, E. M. Cornelisse, J.: ibid. 527 (1985)
95. Osselton, E. M., Eyken, C. P., Jans, A. W. H., Cornelisse, J.: ibid. 1577 (1985)
96. Mattay, J., Runsink, J., Rumbach, T., Ly, C., Gersdorf, J.: J. Am. Chem. Soc. *107*, 2557 (1985);
    Mattay, J., Runsink, J., Gersdorf, J., Rumbach, T., Ly, C.: Helv. Chim. Acta *69*, 442 (1986)
97. Jans, A. W. H., van Arkel, B., van Dijk-Knepper, J. J., Mioch, H., Cornelisse, J.: Tetrahedron *40*, 5071 (1984)
98. Osselton, E. M., Lempers, E. L. M., Cornelisse, J.: Recl. Trav. Chim. Pays-Bas *104*, 124 (1985)
99. Al-Jalal, N., Drew, M. G. B., Gilbert, A.: J. Chem. Soc., Chem. Commun. 85 (1985)
100. Ellis-Davies, G. C. R., Cornelisse, J.: Tetrahedron Lett. 1893 (1985)
101. Ellis-Davies, G. C. R., Gilbert, A., Heath, P., Lane, J. C., Warrington, J. V., Westover, D. L.: J. Chem. Soc. Perkin Trans. II 1833 (1984)
102. deVaal, P., Lodder, G., Cornelisse, J.: Tetrahedron Lett. 4395 (1985)
103. Sheridan, R. S.: J. Am. Chem. Soc. *105*, 5140 (1983)
104. Fray, G. I., Hearn, G. M., Petts, J. C.: Tetrahedron Lett. 2923 (1984)
105. Hafner, K., Thiele, G. F.: J. Am. Chem. Soc. *107*, 5526 (1985)
106. Huisgen, R., Bronberger, F.: Tetrahedron Lett. 61 (1984)
107. Martin, H.-D., Urbanek, T., Pföhler, P., Walsh, R.: J. Chem. Soc., Chem. Commun. 964 (1985)
108. Klumpp, G. W., Schakel, M.: Tetrahedron Lett. 4595 (1983)
109. Piers, E., Banville, J., Lau, C. K., Nagakura, I.: Can. J. Chem. *60*, 6965 (1983)
110. Boaventura, M. A., Drouin, J., Conia, J. M: Synthesis 801 (1983)
111. Ziegler, F. E., Mikami, K.: Tetrahedron Lett. 127 (1984)
112. Ziegler, F. E., Mencel, J. J.: ibid. 123 (1984)
113. Mehta, G., Srikrishna, A., Suri, S. C., Nair, M. S.: J. Org. Chem. *48*, 5107 (1983)
114. Mehta, G., Rao, K. S., Marchand, A. P., Kaya, R.: ibid. *49*, 3848 (1984)

115. Okamoto, Y., Kanematsu, K., Fujiyoshi, T., Osawa, E.: Tetrahedron Lett. 5645 (1983)
116. Okamoto, Y., Harano, K., Yasuda, M., Osawa, E., Kanematsu, K.: Chem. Pharm. Bull. *31*, 2526 (1983);
    Okamoto, Y., Senokuchi, K., Kanematsu, K.: ibid. *32*, 4593 (1984)
117. Mehta, G., Reddy, D. S., Reddy, A. V.: Tetrahedron Lett. 2275 (1984)
118. Ogino, T., Awano, K., Ogihara, T., Isogai, K.: ibid. 2781 (1983)
119. Mehta, G., Nair, M. S.: J. Chem. Soc., Chem. Commun. 629 (1985)
120. Mehta, G., Nair, M. S.: J. Am. Chem. Soc. *107*, 7519 (1985)
121. Paquette, L. A., Nakamura, K., Fischer, J. W.: Tetrahedron Lett. 4051 (1985)
122. Mehta, G., Rao, K. S.: ibid. 809 (1983)
123. Mehta, G., Rao, K. S.: J. Org. Chem. *50*, 5537 (1985)
124. Klunder, A. J. H., Ariaans, G. J. A., Zwanenburg, B.: Tetrahedron Lett. 5457 (1984);
    Klunder, A. J. H., Ariaans, G. J. A., van der Loop, E. A. R. M., Zwanenburg, B.: Tetrahedron *42*, 1903 (1986)
125. Wilt, J. W., Curtis, V. A., Congson, L. N., Palmer, R.: J. Org. Chem. *49*, 2937 (1984)
126. Finkelshtein, E. S., Eremeishvili, M. G., Yatsenko, M. S., Portnykh, E. B., Vdonin, V. M.: Neftekhimiya *25*, 48 (1985)
127. Kienzle, F., Minder, R. E.: Chimia *39*, 100 (1985)
128. Fitjer, L., Wehle, D., Noltemeyer, M., Egert, E., Sheldrick, G. M: Chem. Ber. *117*, 203 (1984)
129. Fitjer, L., Kühn, W., Klages, U., Egert, E., Clegg, W., Schormann, N., Sheldrick, G. M.: ibid. *117*, 3075 (1984)
130. Fitjer, L., Kanschik, A., Majewski, M.: Tetrahedron Lett. 5277 (1985)
131. Kakiuchi, K., Kumanoya, S., Ue, M., Tobe, Y., Odaira, Y.: Chem. Lett. 989 (1985)
132. Kakiuchi, K., Ue, M., Wakake, I., Tobe, Y., Odaira, Y., Yasuda, M., Shima, K.: J. Org. Chem. *51*, 281 (1986)
133. Smith, A. B., III, Wexler, B. A.: Tetrahedron Lett. 2317 (1984);
    Smith, A. B., III, Wexler, B. A., Tu, C.-Y., Konopelski, J. P.: J. Am. Chem. Soc, *107*, 1308 (1985)
134. Kakiuchi, K., Itoga, K., Tsugaru, T., Hato, Y., Tobe, Y., Odaira, Y.: J. Org. Chem. *49*, 659 (1984)
135. Mehta, G., Pramod, K.; Subrahmanyam, D.: J. Chem. Soc., Chem. Commun. 247 (1986)
136. Klester, A. M., Ganter, C.: Helv. Chim. Acta *68*, 104 (1985);
    Klester, A. M., Ganter, C.: ibid. *68*, 734 (1985)
137. McAndrew, B. A., Meakins, S. E., Sell, C. S., Brown, C.: J. Chem. Soc. Perkin Trans. I 1373 (1983)
138. Khomenko, T. M., Bagryanskaya, I. Y., Gatilov, Y. V., Korchagina, D. V., Gatilova, V. P., Dubovenko, Z. V., Barkhash, V. A.: Zh. Org. Khim. *21*, 677 (1985)
139. Shitole, H. R., Vyas, P., Nayak, U. R.: Tetrahedron Lett. 2411 (1983)
140. Duc, D. K. M., Fetizon, M., Kone, M.: Tetrahedron *34*, 3513 (1978)
141. Murata, Y., Ohtsuka, T., Shirahama, H., Matsumoto, T.: Tetrahedron Lett. 4313 (1981)
142. Inoue, Y., Daino, T., Hagiwara, S., Nakamura, H., Hakushi, T.: J. Chem. Soc., Chem. Commun. 804 (1985)
143. Kulagowski, J. J., Moody, C. J., Rees, C. W.: J. Chem. Soc. Perkin Trans. I 2725 (1985)
144. Dauben, W. G., Shapiro, G., Luders, L.: Tetrahedron Lett. 1429 (1985)

145. Lehr, K.-H., Hildebrand, R., Fritz, H., Knothe, L., Krüger, C., Prinzbach, H.: Chem. Ber. *115*, 1875 (1982)
146. Lehr, K.-H., Hunkler, D., Hädicke, E., Prinzbach, H.: ibid. *115*, 1857 (1982)
147. Feldman, K. S., Come, J. H., Freyer, A. J., Kosmider, B. J., Smith, C. M.: J. Am. Chem. Soc. *108*, 1327 (1986)
148. Lyle, T. A., Mereyala, H. B., Pascual, A., Frei, B.: Helv. Chim. Acta *67*, 774 (1984)
149. Jagodzinski, J. J., Sicinski, R. R.: Tetrahedron Lett. 3901 (1981)
150. Fox, M. A., Singletary, N. J.: J. Org. Chem. *47*, 3412 (1982)
151. Sustmann, R., Dern, H.-J.: Chem. Ber. *116*, 2958 (1983)
152. Jurlina, J. L., Patel, H. A., Stothers, J. B.: Can. J. Chem. *62*, 1159 (1984)
153. Patel, H. A., Stothers, J. B.: ibid. *62*, 1926 (1984)
154. Ragauskas, A. J., Stothers, J. B.: ibid. *63*, 1250 (1985)
155. Fleischhauer, I., Brinker, U. H.: Tetrahedron Lett. 3205 (1983)
156. Brinker, U. H., König, L.: Chem. Lett. 45 (1984)
157. Kirmse, W., Ritzer, J.: Chem. Ber. *118*, 4987 (1985)
158. Kusuda, K.: Bull. Chem. Soc. Jpn. *56*, 481 (1983)
159. Clark, G. R., Thiensathit, S.: Tetrahedron Lett. 2503 (1985)
160. Little, R. D., Stone, K. J.: J. Am. Chem. Soc. *105*, 6976 (1983); Stone, K. J., Little, R. D.: ibid. *107*, 2495 (1985)
161. Little, R. D., Moeller, K. D.: J. Org. Chem. *48*, 4487 (1983)
162. Campopiano, I., Little, R. D., Petersen, J. L.: J. Am. Chem. Soc. *107*, 3721 (1985)
163. Sabatelli, A. D., Salinaro, R. F., Mondo, J. A., Berson, J. A.: Tetrahedron Lett. 5851 (1985)
164. Little, R. D., Bode, H., Stone, K. J., Wallquist, O., Dannecker, R.: J. Org. Chem. *50*, 2400 (1985)
165. Moeller, K. D., Little, R. D.: Tetrahedron Lett. 3417 (1985)
166. Hanold, N., Molz, T., Meier, H.: Angew. Chem., Int. Ed. Engl. *21*, 917 (1982)
167. Nifantev, E. E., Maslennikova, V. I., Magdeeva, R. K.: Zh. Obshch. Khim. *54*, 2349 (1984)
168. Uemura, S., Fukuzawa, S., Toshimitsu, A., Okano, M.: J. Org. Chem. *48*, 270 (1983)
169. Zefirov, N. S., Zyk, N. V.; Kolbasenko, S. I., Kutateladze, A. G.: ibid. *50*, 4539 (1985)
170. Gancarz, R. A., Kice, J. L.: ibid. *46*, 4899 (1981)
171. Wiberg, K. B., Matturo, M. G., Okarma, P. J., Jason, M. E.: J. Am. Chem. Soc. *106*, 2194 (1984)
172. Wiberg, K. B., Adams, R. D., Okarma, P. J., Matturo, M. G., Segmuller, B.: ibid. *106*, 2200 (1984)
173. Ashby, E. C., Wenderoth, B., Pham, T. N., Park, W.-S.: J. Org. Chem. *49*, 4505 (1984)
174. Apparu, M., Barrelle, M.: Bull. Soc. Chim. France 156 (1984)
175. Okarma, P. J., Caringi, J. J.: Org. Prep. Proced. Int. *17*, 212 (1985)
176. Pauw, J. E., Weedon, A. C.: Tetrahedron Lett. 5485 (1982)
177. Lorenzi-Riatsch, A., Nakashita, Y., Hesse, M.: Helv. Chim. Acta *67*, 249 (1984)
178. Mehta, G., Rao, K. S.: Tetrahedron Lett. 3481 (1984)
179. Camps, R., Figueredo, M.: Can. J. Chem. *62*, 1184 (1984)
180. Cohen, E., Ginsburg, D.: Tetrahedron *40*, 5273 (1984)
181. Senegör, L., Ashkenazi, P., Ginsburg, D.: ibid. *40*, 5271 (1984)
182. Crisp, G. T., Scott, W. J.: Synthesis 335 (1985)
183. Ladlow, M., Pattenden, G.: Synth. Commun. *14*, 11 (1984)
184. Trost, B. M., Mao, M. K.-T.: J. Am. Chem. Soc. *105*, 6753 (1983)
185. Trost, B. M., Brandi, A.: ibid. *106*, 5041 (1984)

186. Hellou, J., Kingston, J. F., Fallis, A. G.: Synthesis 1014 (1984)
187. Bunce, R. A., Schlecht, M. F., Dauben, W. G., Heathcock, C. H.: Tetrahedron Lett. 4943 (1983)
188. Trost, B. M., Self, C. R.: J. Org. Chem. *49*, 468 (1984)
189. Bertz, S. H.: Tetrahedron Lett. 5577 (1983)
190. Belletire, J. L., Adams, K. G.: ibid. 5575 (1983)
191. Paquette, L. A., Lau, C. J.: Synth. Commun. *14*, 1081 (1984)
192. Carceller, E., Moyano, A., Serratosa, F.: Tetrahedron Lett. 2031 (1984)
193. Jähne, G., Gleiter, R.: Angew. Chem., Int. Ed. Engl. *22*, 488 (1983)
194. Ikeda, M., Takahashi, M.; Uchino, T., Ohno, K., Tamura, Y., Kido, M.: J. Org. Chem. *48*, 4241 (1983)
195. Mongelli, N., Andreoni, A., Zuliani, L., Gandolfi, C. A.: Tetrahedron Lett. 3527 (1983)
196. Mikhail, G., Demuth, M.: Helv. Chim. Acta *66*, 2362 (1983)
197. Murata, I., Sugihara, Y., Sugimura, T., Wakabayashi, S.: Tetrahedron *42*, 1745 (1986)
198. Mehta, G., Veera Reddy, A., Srikrishna, A.: J. Chem. Soc. Perkin Trans. I 291 (1986)
199. Rao, R. R., Chakraborti, K., Bhattacharya, S., Rao, A.: Ind. J. Chem. *22 B*, 1122 (1983)
200. Rao, R. R., Bhattacharya, S.: ibid. *21 B*, 405 (1982)
201. Brown, H. C., Vander Jagt, D. L., Rothberg, I., Hammar, W. J., Kawakami, J. H.: J. Org. Chem. *50*, 2179 (1985)
202. Brown, H. C., Rothberg, I., Chandrasekharan, J.: ibid. *50*, 5574 (1985)
203. Nakazaki, M., Naemura, K., Hashimoto, M.: ibid. *48*, 2289 (1983)
204. Glanzmann, M., Schröder, G.: Chem. Ber. *116*, 2903 (1983)
205. Paquette, L. A., DeLucca, G., Ohkata, K., Gallucci, J.: J. Am. Chem. Soc. *107*, 1015 (1985)
206. Paquette, L. A., DeLucca, G., Korp, J. D., Bernal, I., Swartzendruber, J. K., Jones, N. D.: ibid. *106*, 1122 (1984)
207. Paquette, L. A., Williams, R. V., Vazeux, M., Browne, A. R.: J. Org. Chem. *49*, 2194 (1984)
208. Drouin, J., Rousseau, G.: J. Organomet. Chem. *289*, 223 (1985)
209. Salomon, M. F., Pardo, S. N., Salomon, R. G.: J. Org. Chem. *49*, 2446 (1984)
    Salomon, M. F., Pardo, S. N., Salomon, R. G.: J. Am. Chem. Soc. *106*, 3797 (1984)
210. Whitesell, J. K., Allen, D. E.: J. Org. Chem. *50*, 3026 (1985)
211. Carceller, E., Castello, A., Garcia, M. L., Moyano, A., Serratosa, F.: Chem. Lett. 775 (1984)
212. Carceller, E., Garcia, M. L., Moyano, A., Serratosa, F.: J. Chem. Soc., Chem. Commun. 825 (1984)
213. Mash, E. A., Nelson, K. A.: Tetrahedron Lett. 1441 (1986)
214. Mehta, G., Narayani Murty, A.: J. Chem. Soc., Chem. Commun. 1058 (1984)
215. Farnum, D. G., Monego, T.: Tetrahedron Lett. 1361 (1983)
216. Burger, U., Bianco, B.: Helv. Chim. Acta *66*, 60 (1983)
217. Trost, B. M., Masuyama, Y.: Tetrahedron Lett. 173 (1984)
218. Whitesell, J. K., Minton, M. A., Felman, S. W.: J. Org. Chem. *48*, 2193 (1983)
219. Wilhelm, D., Clark, T., von R. Schleyer, P., Davies, A. G.: J. Chem. Soc., Chem. Commun. 558 (1984)
220. Askani, R., Hornykiewytsch, T., Müller, K. M.: Tetrahedron Lett. 5513 (1983)
221. Sheridan, R. S.: J. Am. Chem. Soc. *105*, 5140 (1983)
222. Bertz, S. H., Dabbagh, G., Cook, J. M., Honkan, V.: J. Org. Chem. *49*, 1739 (1984)

223. Nakatani, K., Isoe, S.: Tetrahedron Lett. 5335 (1984)
224. Gardette, D., Gramain, J.-C., Lhomme, J., Pascard, C., Prangé, T.: Bull. Soc. Chim. Fr. II–404 (1984)
225. Ermer, O., Mason, S. A.: J. Chem. Soc., Chem. Commun. 53 (1983)
226. Ermer, O., Bödecker, C.-D., Prent, H.: Angew. Chem., Int. Ed. Engl. 23, 55 (1984)
227. Gallucci, J. C., Kravetz, T. M., Green, K. E., Paquette, L. A.: J. Am. Chem. Soc. 107, 6592 (1985)
228. Watson, W. H., Galloy, J., Bartlett, P. D., Roof, A. A. M.: ibid. 103, 2022 (1981)
229. Hagenbuch, J.-P., Vogel, P., Pinkerton, A. A., Schwarzenbach, D.: Helv. Chim. Acta 64, 1818 (1981)
230. Paquette, L. A., Charumilind, P., Böhm, M. C., Gleiter, R., Bass, L. S., Clardy, J.: J. Am. Chem. Soc. 105, 3136 (1983)
231. Paquette, L. A., Hayes, P. C., Charumilind, P., Böhm, M. C., Gleiter, R., Blount, J. F.: ibid. 105, 3148 (1983)
232. Pinkerton, A. A., Schwarzenbach, D., Stibbard, J. H., Carrupt, P. A., Vogel, P.: ibid. 103, 2095 (1981)
233. Paquette, L. A., Schaefer, A. G., Blount, J. F.: ibid. 105, 3642 (1983)
234. Mackenzie, K., Miller, A. S., Muir, K. W., Manojlovic-Muir, Lj.: Tetrahedron Lett. 4747 (1983)
235. Ermer, O., Bödecker, C. D.: Helv. Chim. Acta 66, 943 (1983)
236. Paquette, L. A., Green, K. E., Hsu, L.-Y.: J. Org. Chem. 49, 3650 (1984)
237. Gajhede, M., Jørgensen, F. S., Kopecky, K. R., Watson, W. H., Kashyap, R. P.: ibid. 50, 4395 (1985)
238. Paquette, L. A., Hsu, L.-Y., Gallucci, J. C., Korp, J. D., Bernal, I., Kravetz, T. M., Hathaway, S. J.: J. Am. Chem. Soc. 106, 5743 (1984)
239. Paquette, L. A., Hathaway, S. J., Kravetz, T. M., Hsu, L.-Y.: ibid. 106, 5741 (1984)
240. Paquette, L. A., Hathaway, S. J., Schirch, P. F. T.: J. Org. Chem. 50, 4199 (1985)
241. Paquette, L. A., Kravetz, T. M.: ibid. 50, 3781 (1985)
242. Chow, T. J., Liu, L.-K., Chao, Y.-S.: J. Chem. Soc., Chem. Commun. 700 (1985)
243. Engel, P., Fischer, J. W., Paquette, L. A.: Z. Kristallogr. 166, 225 (1984)
244. Gallucci, J. C., Doecke, C. W., Paquette, L. A.: J. Am. Chem. Soc. 108, 1343 (1986)
245. Brown, R. S., Buschek, J. M., Kopecky, K. R. Miller, A. J.: J. Org. Chem. 48, 3692 (1983)
246. Gleiter, R., Jähne, G.: Tetrahedron Lett. 5063 (1983)
247. Gleiter, R., Jähne, G., Müller, G., Nixdorf, M., Irngartinger, H.: Helv. Chim. Acta 69, 71 (1986)
248. Bischof, P., Gleiter, R., Haider, R., Rees, C. W.: J. Chem. Soc. Perkin Trans. II 1001 (1985)
249. Glidewell, C.: ibid. II 1175 (1984)
250. Gleiter, R., Spanget-Larsen, J.: Tetrahedron Lett. 927 (1982); Tetrahedron 39, 3345 (1981)
251. Jørgensen, F. S.: Tetrahedron Lett. 5289 (1983)
252. Houk, K. N., Rondan, N. G., Brown, F. K., Jørgensen, W. L., Madura, J. D., Spellmeyer, D. C.: J. Am. Chem. Soc. 105, 5980 (1983)
253. Johnson, C. A.: J. Chem. Soc., Chem. Commun. 1135 (1983)
254. Schulman, J. M., Disch, R. L.: Tetrahedron Lett. 5647 (1985)
255. Schippers, P. H., Dekkers, H. P. J. M.: J. Am. Chem. Soc. 105, 145 (1983)
256. Joseph-Nathan, P., Santillan, R. L., Gutierrez, A.: J. Nat. Prod. 47, 924 (1984)
257. Kataoka, M., Ohmae, T., Nakajima, T.: J. Org. Chem. 51, 358 (1986)
258. Kuwajima, S.: J. Am. Chem. Soc. 106, 6496 (1984)

259. Dewar, M. J. S., Merz, K. M., Jr.: ibid. *107*, 6175 (1985)
260. Stezowski, J. J., Hoier, H., Wilhelm, D., Clark, T., von R. Schleyer, P.: J. Chem. Soc., Chem. Commun. 1263 (1985)
261. Wilhelm, D., Courtneidge, J. L., Clark, T., Davies, A. G.: ibid. 810 (1984)
262. Hafner, K., Goltz, M.: Angew. Chem. *94*, 711 (1982)
263. Askani, R., Kalinowski, H.-O., Weuste, B.: Org. Magn. Reson. *18*, 176 (1982)
264. Askani, R., Kalinowski, H.-O., Pelech, B., Weuste, B.: Tetrahedron Lett. 2321 (1984)
265. Schnieders, C., Müllen, K., Braig, C., Schuster, H., Sauer, J.: ibid. 749 (1984)
266. Sellner, I., Schuster, H., Sichert, H., Sauer, J., Nöth, H.: Chem. Ber. *116*, 3751 (1983)
267. Quast, H., Görlach, Y.: Tetrahedron Lett. 5591 (1983)
268. Quast, H., Christ, J., Peters, E.-M., Peters, K., von Schnering, H. G.: Chem. Ber. *118*, 1154 (1985)
269. Quast, H., Görlach, Y., Christ, J., Peters, E.-M., Peters, K., von Schnering, H. G., Jackman, L. M., Ibar, G., Freyer, A. J.: Tetrahedron Lett. 5595 (1983)
270. Quast, H., Christ, J.: Liebigs Ann. Chem. 1180 (1984)
271. Quast, H., Christ, J., Peters, E.-M., Peters, K., von Schnering, H. G.: Chem. Ber. *118*, 1176 (1985)
272. Quast, H., Christ, J.: Angew. Chem., Int. Ed. Eng. *23*, 631 (1984)
273. Miller, L. S., Grohmann, K., Dannenberg, J.: J. Am. Chem. Soc. *105*, 6862 (1983)
274. Askani, R., Littmann, M.: Tetrahedron Lett. 5519 (1985)
275. Martin, H.-D., Urbanek, T., Walsh, R.: J. Am. Chem. Soc. *107*, 5532 (1985)
276. Goldstein, M. J., Wenzel, T. T.: J. Chem. Soc., Chem. Commun. 1655 (1984)
277. Kopecky, K. R., Miller, A. J.: J. Can. Chem. *62*, 1840 (1984)
278. Paquette, L. A., Künzer, H., Green, K. E.: J. Am. Chem. Soc. *107*, 4788 (1985)
279. De Lucchi, O., Licini, G., Pasquato, L.: J. Chem. Soc., Chem. Commun. 418 (1985)
280. Paquette, L. A., Künzer, H., Green, K. E., DeLucchi, O., Licini, G., Pasquato, L., Valle, G.: J. Am. Chem. Soc. *108*, 3453 (1986)
281. Brown, F. K., Houk, K. N.: ibid. *107*, 1971 (1985)
282. Hathaway, S. J., Paquette, L. A.: Tetrahedron *41*, 2037 (1985)
283. Paquette, L. A., Green, K. E., Hsu, L.-Y.: J. Org. Chem. *49*, 3650 (1984)
284. Paquette, L. A., Kravetz, T. M., Hsu, L.-Y.: J. Am. Chem. Soc. *107*, 6598 (1985)
285. Paquette, L. A., Green, K. E., Gleiter, R., Schäfer, W., Gallucci, J. C.: ibid. *106*, 8232 (1984)
286. Charumilind, P., Paquette, L. A.: ibid. *106*, 8225 (1984)
287. Paquette, L. A., Hathaway, S. J., Kravetz, T. M., Hsu, L.-Y.: ibid. *106*, 5741 (1984)
288. Paquette, L. A., Hathaway, S. J., Gallucci, J. G.: Tetrahedron Lett. 2659 (1984)
289. Paquette, L. A., Hathaway, S. J., Schirch, P. F. T.: J. Org. Chem. *50*, 4199 (1985)
290. Paquette, L. A., Kravetz, T. M.: ibid. *50*, 3781 (1985)
291. Paquette, L. A., Charumilind, P., Gallucci, J. C.: J. Am. Chem. Soc. *105*, 7364 (1983)
292. Bartlett, P. D., Wu, C.: J. Org. Chem. *49*, 1880 (1984)
293. Bartlett, P. D., Wu, C.: ibid. *50*, 4087 (1985)
294. Wu, C., Bartlett, P. D.: ibid. *50*, 733 (1985)
295. Bartlett, P. D., Roof, A. A. M., Subramanyam, R., Winter, W. J.: ibid. *49*, 1875 (1984)
296. Roof, A. A. M., Winter, W. J., Bartlett, P. D.: ibid. *50*, 4093 (1985)
297. Bartlett, P. D., Combs, G. L., Jr.: ibid. *49*, 625 (1984)
298. Bartlett, P. D., Ghosh, T.: Tetrahedron Lett. 2613 (1985)

299. Bartlett, P. D., Roof, A. A. M., Winter, W. J.: J. Am. Chem. Soc. *103*, 6520 (1981)
300. Bertz, S. H., Lannoye, G., Cook, J. M.: Tetrahedron Lett. 4695 (1985)
301. Carceller, E., Garcia, M. L., Moyano, A., Pericas, M. A., Serratosa, F.: Tetrahedron *41*, 1831 (1986)
302. Paquette, L. A., Kramer, J. D.: J. Org. Chem. *49*, 1445 (1984)
303. Butenschön, H., de Meijere, A.: Tetrahedron Lett. 4563 (1983); Chem. Ber. *118*, 2557 (1985)
304. Butenschön, H., de Meijere, A.: Tetrahedron Lett. 1693 (1984); Helv. Chim. Acta *68*, 1658 (1985)
305. Butenschön, H., de Meijere, A.: Angew. Chem., Int. Ed. Engl. *23*, 707 (1984); Tetrahedron *41*, 1721 (1986)
306. Gilchrist, T. L., Rees, C. W., Tuddenham, D.: J. Chem. Soc. Perkin Trans. I 83 (1983)
307. McCague, R., Moody, C. J., Rees, C. W.: ibid. I 165 (1984)
308. Gibbard, H. C., Moody, C. J., Rees, C. W.: ibid. I 731 (1985)
309. Gibbard, H. C., Moody, C. J., Rees, C. W.: ibid. I 735 (1985)
310. McCague, R., Moody, C. J., Rees, C. W., Williams, D. J.: ibid. I 909 (1984)
311. McCague, R., Moody, C. J., Rees, C. W.: ibid. I, 2399 (1983)
312. Lidert, Z., McCague, R., Moody, C. J., Rees, C. W.: ibid. I, 383 (1985)
313. Lidert, Z., Rees, C. W.: J. Chem. Soc., Chem. Commun. 317 (1983)
314. Gibbard, H. C., Moody, C. J., Rees, C. W.: J. Chem. Soc. Perkin Trans. I 723 (1985)
315. McCague, R., Moody, C. J., Rees, C. W.: ibid. I, 175 (1984)
316. Mani, J., Cho, J.-H., Astik, R. R., Stamm, E., Bigler, P., Mayer, V., Keese, R.: Helv. Chim. Acta *67*, 1930 (1984)
317. Luyten, M., Keese, R.: Angew. Chem., Int. Ed. Engl. *23*, 390 (1984)
318. Luyten, M., Keese, R.: Helv. Chim. Acta *67*, 2242 (1984)
319. Luyten, M., Keese, R.: Tetrahedron *42*, 1687 (1986)
320. Rao, V. B., Wolff, S., Agosta, W. C.: J. Chem. Soc., Chem. Commun. 293 (1984)
321. Rao, V. B., George, C. F., Wolff, S., Agosta, W. C.: J. Am. Chem. Soc. *107*, 5732 (1985)
322. Rao, V. B., Wolff, S., Agosta, W. C.: Tetrahedron *42*, 1549 (1986)
323. Mani, J., Keese, R.: ibid. *41*, 5697 (1985)
324. Pfenninger, A., Roesle, A., Keese, R.: Helv. Chim. Acta *68*, 493 (1985)
325. Crimmins, M. T., Mascarella, S. W., Bredon, L. D.: Tetrahedron Lett. 997 (1985)
326. Deshpande, M. N., Jawdosiuk, M., Kubiak, G., Venkatachalam, M., Weiss, U., Cook, J. M.: J. Am. Chem. Soc. *107*, 4786 (1985)
327. Venkatachalam, M., Deshpande, M. N., Jawdosiuk, M., Kubiak, G., Wehrli, S., Cook, J. M., Weiss, U.: Tetrahedron *42*, 1597 (1986)
328. Böhm, M. C., Gleiter, R., Schang, P.: Tetrahedron Lett. 2575 (1979)
329. Venkatachalam, M., Jawdosiuk, M., Deshpande, M., Cook, J. M.: ibid. 2275 (1985)
330. Venkatachalam, M., Kubiak, G., Cook, J. M., Weiss, U.: ibid. 4863 (1985)
331. Nakazaki, M., Naemura, K., Hashimoto, M.: Bull. Chem. Soc. Jpn. *56*, 2543 (1983)
332. Naemura, K., Katoh, T., Fukunaga, R., Chikamatsu, H., Nakazaki, M.: ibid. *58*, 1407 (1985)
333. Nakazaki, M., Naemura, K., Harada, H., Narukaki, H.: J. Org. Chem. *47*, 3479 (1982)
334. Fessner, W.-D., Prinzbach, M.: Tetrahedron *42*, 1797 (1986)
335. Naemura, K., Hokura, Y., Nakazaki, M.: ibid. *42*, 1763 (1986)

336. Majerski, Z., Hamersak, Z., Mlinaric-Majerski, K.: J. Chem. Soc., Chem. Commun. 1830 (1985)
337. Garratt, P. J., Doecke, C. W., Weber, J. C., Paquette, L. A.: J. Org. Chem. *51*, 449 (1986)
338. Paquette, L. A., Fischer, J. W., Browne, A. R., Doecke, C. W.: J. Am. Chem. Soc. *107*, 686 (1986)
339. Eaton, P. E., Srikrishna, A., Uggieri, F.: J. Org. Chem. *49*, 1728 (1984)
340. Eaton, P. E., Bunnelle, W. H.: Tetrahedron Lett. 23 (1984);
     Eaton, P. E., Bunnelle, W. H., Engel, P.: Can. J. Chem. *62*, 2612 (1984)
341. Mehta, G., Rao, H. S. P.: J. Chem. Soc., Chem. Commun. 472 (1986)
342. Mehta, G., Nair, M. S.: ibid. 439 (1986)
343. Baldwin, J. E., Beckwith, P. L. M.: ibid. 279 (1983)
344. Baldwin, J. E., Beckwith, P. L. M., Wallis, J. D., Orrell, A. P. K., Prout, K.: J. Chem. Soc. Perkin Trans. II 53 (1984)
345. Fessner, W.-D., Prinzbach, H., Rihs, G.: Tetrahedron Lett. 5857 (1983)
346. Paquette, L. A., Miyahara, Y., Doecke, C. W.: J. Am. Chem. Soc. *108*, 1716 (1986)
347. Kaiser, R., Lamparsky, D.: Helv. Chim. Acta *66*, 1843 (1983)
348. Forster, P. G., Ghisalberti, E. L., Jefferies, P. R., Skelton. B. W., White, A. H.: Tetrahedron *42*, 215 (1986)
349. Subramaniam, G. B. V., Rizoi, R. N., Iqbal, J., Chander, Y.: Ind. J. Chem. *22 B*, 831 (1983)
350. Kaisin, M., Tursch, B., Declercq, J. P., Germain, G., Van Meerssche, M.: Bull. Soc. Chim. Belg. *88*, 253 (1979)
351. Kaisin, M., Braekman, J. C., Daloze, D., Tursch, B.: Tetrahedron *41*, 1067 (1985)
352. Nakagawa, M., Hirota, A., Sakai, H., Isogai, A.: J. Antibiot. *35*, 778 (1982)
353. Hirota, A., Nakagawa, M., Sakai, H., Isogai, A.: ibid. *35*, 783 (1982)
354. Hirota, A., Nakagawa, M., Sakai, H., Isogai, A.: ibid. *37*, 475 (1984)
355. Hirota, A., Nakagawa, M., Hirota, H., Takahashi, T., Isogai, A.: ibid. *39*, 149 (1986)
356. Tillman, A. M., Cane, D. E.: ibid. *36*, 170 (1983)
357. Sato, H., Sasaki, T., Yonehara, H., Takahashi, S., Takeuchi, M., Kuwano, H., Arai, M.: ibid. *37*, 1076 (1984)
358. Groweiss, A., Fenical, W., He, C., Clardy, J., Wu, Z., Yiao, Z., Long, K.: Tetrahedron Lett. 2379 (1985)
359. Anke, T., Heim, J., Knoch, F., Mocek, U., Steffan, B., Steglich, W.: Angew. Chem., Int. Ed. Engl. *24*, 709 (1985)
360. Bohlmann, F., Zdero, C.: Phytochem. *21*, 139 (1982)
• 361. Bohlmann, F., Ziesche, J., Gupta, R. K.: ibid. *21*, 1331 (1982)
362. Bohlmann, F., Zdero, C., King, R. M., Robinson, H.: ibid. *22*, 1201 (1983)
363. Bohlmann, F., Misra, L. N., Jakupovic, J., Robinson, H., King, R. M.: J. Nat. Prod. *47*, 658 (1984)
364. Bohlmann, F., Misra, L. N., Jakupovic, J.: Phytochem. *24*, 1378 (1985)
365. Ruest, L., Taylor, D. R., Deslongchamps, P.: Can. J. Chem. *63*, 2840 (1985)
366. Cane, D. E., Tillman, A. M.: J. Am. Chem. Soc. *105*, 122 (1983)
367. Seto, H., Noguchi, H., Sankawa, U., Titaka, Y.: J. Antibiot. *37*, 816 (1984)
368. Cane, D. E., Whittle, Y. G., Liang, T.-C.: Tetrahedron Lett. 1119 (1984)
369. Hirota, A., Nakagawa, M., Sakai, H., Isogai, A., Furihata, K., Seto, H.: ibid. 3845 (1985)
370. Beale, J. M., Jr., Chapman, R. L. Rosazza, J. P. N.: J. Antibiot. *37*, 1376 (1984)
371. Nakagawa, A., Tomoda, H., Hao, M. V., Okano, K., Iwai, Y., Omura, S.: ibid. *38*, 1114 (1985)
372. Hanton, L. R., Simpson, J., Weavers, R. T.: Austr. J. Chem. *36*, 2581 (1983)

373. Corbett, R. E., Jogia, M. K., Lauren, D. R., Weavers, R. T.: ibid. *36*, 1001 (1983)
374. Demuth, M., Schaffner, K.: Angew. Chem., Int. Ed. Engl. *24*, 820 (1982); Demuth, M.: Chimia *38*, 257 (1984)
375. Abe, F., Mori, T., Yamanuchi, Y.: Chem. Pharm. Bull. *32*, 2947 (1984)
376. Parkes, K. E. B., Pattenden, G.: Tetrahedron Lett. 1305 (1986)
377. Trost, B. M., Balkovec, J. M., Mao, M. K.-T.: J. Am. Chem. Soc. *105*, 6755 (1983)
378. Trost, B. M., Balkovec, J. M.: Tetrahedron Lett. 1807 (1985)
379. Marini-Bettolo, G. B., Nicoletti, M., Messana, I., Patamia, M., Galeffi, C., Oguakwa, J. U., Portalone, G., Vaciago, A.: Tetrahedron *39*, 323 (1983)
380. Lee, T. V., Toczek, J., Roberts, S. M.: J. Chem. Soc., Chem. Commun. 371 (1985)
381. Nakatani, K., Isoe, S.: Tetrahedron Lett. 2209 (1985)
382. Connolly, P. J., Heathcock, C. H.: J. Org. Chem. *50*, 4135 (1985)
383. Inoue, H., Nishioka, T.: Chem. Pharm. Bull. (Japan) *21*, 497 (1973)
384. Takemoto, T., Isoe, S.: Chem. Lett. 1931 (1982)
385. Boeckman, R. K., Jr., Napier, J. J., Thomas, E. W., Sato, R. I.: J. Org. Chem. *48*, 4153 (1983)
386. Ritterskamp, P., Demuth, M., Schaffner, K.: ibid. *49*, 1155 (1984)
387. Wender, P. A., Dreyer, G. B.: Tetrahedron Lett. 4543 (1983)
388. Au-Yeung, B., Fleming, I.: J. Chem. Soc., Chem. Commun. 81 (1977)
389. Kon, K., Isoe, S.: Tetrahedron Lett. 3399 (1980); Trost, B. M., Chan, D. M. T.: J. Am. Chem. Soc. *103*, 5972 (1981)
390. Kon, K., Isoe, S.: Helv. Chim. Acta *66*, 755 (1983)
391. Büchi, G., Carlson, J. A., Powell, J. E., Jr., Tietze, L. F.: J. Am. Chem. Soc. *92*, 2165 (1970); *Idem., Ibid.* *95*, 540 (1973)
392. Caille, J. C., Bellamy, F., Guilard, R.: Tetrahedron Lett. 2345 (1976)
393. Demuth, M., Chandrasakhar, S., Schaffner, K.: J. Am. Chem. Soc. *106*, 1092 (1984)
394. Trost, B. M., Nanninga, T. N.: ibid. *107*, 1293 (1985)
395. Storme, P., Callant, P., Vandewalle, M.: Bull. Soc. Chim. Belg. *92*, 1019 (1983)
396. Piers, E., Gavai, A. V.: Tetrahedron Lett., 313 (1986)
397. Harding, K. E., Clement, K. S.: J. Org. Chem. *49*, 3870 (1984)
398. Nakatsu, T., Ravi, B. N., Faulkner, D. J.: ibid. *46*, 2435 (1981)
399. Whitesell, J. K., Fisher, M., Da Silva Jardine, P.: ibid. *48*, 1556 (1983)
400. Whitesell, J. K., Wang, P. K. S., Aguilar, D. A.: ibid. *48*, 2511 (1983)
401. Rao, R. R.: Ind. J. Chem. Sect. B *17*, 211 (1979)
402. Stork, G., Clarke, F. H.: J. Am. Chem. Soc. *83*, 5654 (1966)
403. Horton, M., Pattenden, G.: Tetrahedron Lett. 2125 (1983)
404. Horton, M., Pattenden, G.: J. Chem. Soc. Perkin I, 811 (1984)
405. Landry, D. W.: Tetrahedron *39*, 2761 (1983)
406. Woodward, R. B., Katz, T. J.: ibid. *5*, 70 (1959)
407. Irie, H., Takeda, S., Yamamura, A., Mizuno, Y., Tomimasu, H., Ashizawa, K., Taga, T.: Chem. Pharm. Bull. *32*, 2886 (1984)
408. Adrianome, M., Delmond, B.: J. Chem. Soc., Chem. Commun. 1203 (1985)
409. Adrianome, M., Delmond, B.: Tetrahedron Lett. 6341 (1985)
410. Martin, D. G., Slomp, G., Mizsak, S., Duchamp, D. J., Chidester, C. G.: ibid. 4901 (1970)
411. Cane, D. E., Tillman, A. M.: J. Am. Chem. Soc. *105*, 122 (1983), ref. 3 a
412. Murata, Y., Ohtsuka, T., Shirahama, H., Matsumoto, T.: Tetrahedron Lett. 4313 (1981)
413. Ohtsuka, T., Shirahama, H., Matsumoto, T.: ibid. 3851 (1983)
414. Ohtsuka, T., Shirahama, H., Matsumoto, T.: Chem. Lett. 1923 (1984)

415. Cane, D. E., Thomas, P. J.: J. Am. Chem. Soc. *106*, 5295 (1984)
416. Greene, A. E., Luche, M.-J., Déprés, J.-P.: ibid. *105*, 2435 (1983)
417. Paquette, L. A., Annis, G. D., Schostarez, H., Blount, J. F.: J. Org. Chem. *46*, 3768 (1981)
418. Danishefsky, S., Hirama, M., Gombatz, K., Harayama, T., Berman, E., Shuda, P. F.: J. Am. Chem. Soc. *100*, 6535 (1978); *101*, 7020 (1979)
419. Taber, D. F., Schuchardt, J. L.: ibid. *107*, 5289 (1985)
420. Ranieri, R. L., Calton, G. J.: Tetrahedron Lett. 499 (1978);
    Calton, G. J., Ranieri, R. L., Espenshade, M. A.: J. Antibiot. *31*, 38 (1978)
421. Takeda, K., Shimono, Y., Yoshii, E.: J. Am. Chem. Soc. *105*, 563 (1983)
422. Danishefsky, S., Vaughan, K., Gadwood, R. C., Tsuzuki, K.: ibid. *103*, 4136 (1981); *102*, 4262 (1980)
423. Schlessinger, R. H., Wood, J. L., Poss, A. J., Nugent, R. A., Parsons, W. H.: J. Org. Chem. *48*, 1146 (1983)
424. Burke, S. D., Murtiashaw, C. W., Saunders, J. O., Dike, M. S.: J. Am. Chem. Soc. *104*, 872 (1982)
425. Burke, S. D., Murtiashaw, C. W., Oplinger, J. A.: Tetrahedron Lett. 2949 (1983)
426. Burke, S. D., Murtiashaw, C. W., Saunders, J. O., Oplinger, J. A., Dike, M. S.: J. Am. Chem. Soc. *106*, 4558 (1984)
427. Dewanckele, J. M., Zutterman, F.; Vandewalle, M.: Tetrahedron *39*, 3235 (1983)
428. Bornack, W. K., Bhagwat, S. S., Ponton, J., Helquist, P.: J. Am. Chem. Soc. *103*, 4647 (1981)
429. Cooper, K., Pattenden, G.: J. Chem. Soc. Perkin I 799 (1984)
430. Oehldrich, J., Cook, J. M., Weiss, U.: Tetrahedron Lett. 4549 (1976)
431. Smith, A. B., Konopelski, J. P.: J. Org. Chem. *49*, 4094 (1984)
432. Piers, E., Moss, N.: Tetrahedron Lett. 2735 (1985)
433. Katiuchi, K., Nakao, T., Takeda, M., Tobe, Y., Odaira, Y.: ibid. 557 (1984)
434. Smith, A. B., Wexler, B. A., Slade, J.: ibid. 1631 (1982)
435. Nakagawa, M., Hirota, A., Sakai, H., Isogai, A.: J. Antibiot. *35*, 778, 783 (1982)
436. Kon, K., Ito, K., Isoe, S.: Tetrahedron Lett. 3739 (1984)
437. Iwata, C., Masayuki, Y., Aoki, S., Suzuki, K., Takahashi, I., Arakawa, H., Imanishi, T., Tanaka, T.: Chem. Pharm. Bull. *33*, 436 (1985)
438. Wender, P. A., Wolanin, D. J.: J. Org. Chem. *50*, 4418 (1985)
439. Takeda, R., Naoki, H., Iwashita, T., Hirose, Y.: Tetrahedron Lett. 5307 (1981)
440. Huguet, J., Karpf, M., Dreiding, A. S.: 4177 (1983)
441. Hochmannová, J., Novotný, L., Herout, V.: Collect. Czech. Chem. Commun. *27*, 2711 (1962);
    Novotný, L., Herout, V.: ibid. *30*, 3579 (1965)
442. Baldwin, J. E., Barden, T. C.: J. Org. Chem. *48*, 625 (1983)
443. Vokiv, K., Samek, Z., Herout, V., Šorm, F.: Tetrahedron Lett. 1665 (1972)
444. Kreiser, W., Janitschke, L.; Sheldrick, W. S.: J. Chem. Soc., Chem. Commun. 269 (1977);
    Kreiser, W., Janitschke, L., Voss, W., Ernst, L., Sheldrick, W. S.: Chem. Ber. *112*, 397 (1979)
445. Baldwin, J. E., Barden, T. C.: J. Am. Chem. Soc. *105*, 6656 (1983)
446. Baldwin, J. E., Barden, T. C., Cianciosi, S. J.: J. Org. Chem. *51*, 1133 (1986)
447. Kreiser, W., Janitschke, L.: Chem. Ber. *112*, 408 (1979)
448. Baldwin, J. E., Barden, T. C.: J. Org. Chem. *46*, 2442 (1981)
449. Hallsworth, A. S., Henbest, H. B., Wrigley, T. I.: J. Chem. Soc. 1969 (1957)
450. Marshall, J. A., Bundy, G. L., Fanta, W. I.: J. Org. Chem. *33*, 3913 (1968)
451. Trost, B. M., Renaut, P.: J. Am. Chem. Soc. *104*, 6668 (1982)

452. Manzardo, G. G. G., Karpf, M., Dreiding, A. S.: Helv. Chim. Acta 66, 627 (1983)
453. Moncada, S., Gryglewski, R., Bunting, S., Vane, J. R.: Nature 263, 663 (1976); Johnson, R. A., Morton, D. R., Kinner, J. H., Gorman, R. R., McGuire, J. C., Sun, F. F., Whittaker, N., Bunting, S., Salmon, J., Moncada, S., Vane, J. R.: Prostaglandins 12, 915 (1976)
454. Moncada, S., Vane, J. R.: J. Med. Chem. 23, 591 (1980)
455. Nickolson, R. C., Town, M. H., Vorbrüggen, H.: Med. Res. Rev. 5, 1 (1985)
456. Shibasaki, M., Torisawa, Y., Ikegami, S.: Tetrahedron Lett. 3493 (1983)
457. Shibasaki, M., Fukasawa, H., Ikegami, S.: ibid. 3497 (1983)
458. Shimoji, K., Konishi, Y., Arai, Y., Hayashi, M., Yamamoto, H.: J. Am. Chem. Soc. 100, 2547 (1978)
459. Iseki, K., Mase, T., Okazaki, T., Shibasaki, M., Ikegami, S.: Chem. Pharm. Bull. 31, 4448 (1983)
460. Iseki, K., Yamazaki, M., Shibasaki, M., Ikegami, S.: Tetrahedron 37, 4411 (1981); Shibasaki, M., Iseki, K., Ikegami, S.: Synth. Commun. 10, 545 (1980); Shibasaki, M., Iseki, K., Ikegami, S.: Chem. Lett. 1299 (1979)
461. Okazaki, T., Shibasaki, M., Ikegami, S.: Chem. Pharm. Bull. 32, 424 (1984)
462. Robins, M. J., Wilson, J. S.: J. Am. Chem. Soc. 103, 932 (1981)
463. Ogawa, Y., Shibasaki, M.: Tetrahedron Lett. 1067 (1984)
464. Sodeoka, M., Shibasaki, M.: Chem. Lett. 579 (1984)
465. Torisawa, Y., Okabe, H., Shibasaki, M., Ikegami, S.: ibid. 1069 (1984)
466. Torisawa, Y., Okabe, H., Ikegami, S.: J. Chem. Soc., Chem. Commun. 1602 (1984)
467. Shibasaki, M., Sodeoka, M., Ogawa, Y.: J. Org. Chem. 49, 4098 (1984)
468. Bartman, W., Beck, G.: Angew. Chem., Int. Ed. Engl. 21, 751 (1982)
469. Shibasaki, M., Sodeoka, M.: Tetrahedron Lett. 3491 (1985)
470. Mase, T., Sodeoka, M.: ibid. 5087 (1984)
471. Bannai, K., Toru, T., Hazato, A., Ōba, T., Tanaka, T., Okamura, N., Watanabe, K., Kurozumi, S.: Chem. Pharm. Bull. 30, 1102 (1982)
472. Aristoff, P. A., Johnson, P. D., Harrison, A. W.: J. Org. Chem. 48, 5341 (1983)
473. Aristoff, P. A.: ibid. 46, 1954 (1981)
474. Schwartz, J., Carr, D. B., Hansen, R. T., Dayrit, F. M.: ibid. 45, 3053 (1980); Hansen, R. T., Carr, D. B., Schwartz, J.: J. Am. Chem. Soc. 100, 2244 (1978)
475. Stezowski, J. J., Flohé, L., Bohlke, H.: J. Chem. Soc., Chem. Commun. 1315 (1983)
476. Flohé, L., Bohlke, H., Frankus, E., Kim, S. M. A., Lintz, W., Loschen, G., Michel, G., Muller, B., Schneider, J., Seipp, U., Vollenberg, W., Wilsmann, K.: Arzneim-Forsch-Drug Res. 33, 1240 (1983)
477. Riefling, B. F., Radunz, H. E.: Tetrahedron Lett. 5487 (1983)
478. Roberts, S. M., Newton, R. F.: Tetrahedron 36, 2163 (1980)
479. Corey, E. J., Arnold, Z., Hutton, J.: Tetrahedron Lett. 307 (1970)
480. Dupin, C., Dupin, J.-F.: Bull. Soc. Chim. Fr. 249 (1970)
481. Riefling, B. F.: Tetrahedron Lett. 2063 (1985)
482. Koyama, K., Kojima, K.: Chem. Pharm. Bull. 32, 2866 (1984)
483. Amemiya, S., Kojima, K., Sakai, K.: ibid. 32, 4746 (1984)
484. Kojima, K., Koyama, K., Amemiya, S.: Tetrahedron 41, 4449 (1985)
485. Kojima, K., Amemiya, S., Koyama, K., Sakai, K.: Chem. Pharm. Bull. 33, 2688 (1985)
486. Kojima, K., Sakai, K.: Tetrahedron Lett. 3743 (1978)
487. Nicolaou, K. C., Sipio, W. J., Magolda, R. L., Seitz, S., Barnett, W. E.: J. Chem. Soc., Chem. Commun. 1067 (1978)
488. Shibasaki, M., Ueda, J., Ikegami, S.: Tetrahedron Lett. 433 (1979)

489. Vorbrüggen, H., Bennua, B.: Synth. Commun. 925 (1985)
490. Bestmann, H. J., Schade, G., Schmid, G.: Angew. Chem. 92, 856 (1980); Angew. Chem., Int. Ed. Engl. 19, 822 (1980)
491. Bennua, B., Dahl, H., Vorbrüggen, H.: Synth. Commun. 41 (1986)
492. Bird, C. W., Butler, H. I., Caton, M. P. L., Coffee, E. C. J., Hardy, C. J., Hart, T. W., Mason, H. J.: Tetrahedron Lett. 4101 (1985)
493. Mori, K., Tsuji, M.: Tetrahedron 42, 435 (1986)
494. Konishi, Y., Kawamura, M., Iguchi, Y., Arai, Y., Hayashi, M.: ibid. 37, 4391 (1981)
495. Shinoda, M., Iseki, K., Oguri, T., Hayashi, Y., Yamada, S., Shibasaki, M.: Tetrahedron Lett. 87 (1986)
496. Crosby, D. G.: Naturally Occurring Insecticides (eds. Jacobson, M., Crosby, D. G.) Dekker, New York, 198 (1971)
497. Wiesner, K.: Adv. Org. Chem. 8, 295 (1972);
    Wiesner, K., Valenta, Z., Findlay, J. A.: Tetrahedron Lett. 221 (1967)
498. Srivastava, S. N., Przybylska, M.: Can. J. Chem. 46, 795 (1968)
499. Deslongchamps, P.: Pure and Appl. Chem. 49, 1329 (1977)
500. Bélanger, A., Berney, D. J. F., Borschberg, H.-J., Brousseau, R., Doutheau, A., Durand, R., Katayama, H., Lapalme, R., Leturc, D. M., Liao, C.-C., MacLachlan, F. N., Maffrand, J.-P., Marazza, F., Martino, R., Moreau, C., Saint-Laurent, L., Saintonge, R., Soucy, P., Ruest, L., Deslongchamps, P.: Can. J. Chem. 57, 3348 (1979)
501. Waterhouse, A. L., Holden, I., Casida, J. E.: J. Chem. Soc., Chem. Commun. 1265 (1984)
502. Waterhouse, A. L., Holden, I., Casida, J. E.: J. Chem. Soc. Perkin Trans. II 1011 (1985)
503. Nozoe, S., Furukawa, J., Sankawa, U., Shibata, S.: Tetrahedron Lett. 195 (1976)
504. Dawson, B. A., Ghosh, A. K., Jurlina, J. L., Stothers, J. B.: J. Chem. Soc., Chem. Commun. 204 (1983)
505. Little, R. D., Muller, G. W.: J. Am. Chem. Soc. 103, 2744 (1981) and references therein
506. Cristol, S. J., Seifert, W. K., Solway, S. B.: ibid. 82, 2351 (1960)
507. Little, R. D., Higby, R. G., Moeller, K. D.: J. Org. Chem. 48, 3139 (1983)
508. Funk, R. L., Bolton, G. L.: ibid. 49, 5022 (1984)
509. Funk, R. L., Bolton, G. L., Daggett, J. U., Hansen, M. M., Horcher, L. H. M.: Tetrahedron 41, 3479 (1985)
510. Magnus, P., Quagliato, D. A., Huffman, J. C.: Organometallics 1, 1240 (1982);
    Magnus, P., Quagliato, D. A.: Ibid. 1, 1243 (1982)
511. Magnus, P., Quagliato, D. A.: J. Org. Chem. 50, 1621 (1985)
512. Curran, D. P., Rakiewicz, D. M.: J. Am. Chem. Soc. 107, 1448 (1985)
513. Curran, D. P., Rakiewicz, D. M.: Tetrahedron 41, 3943 (1985)
514. Hua, D. H., Sinai-Zingde, G., Venkataraman, S.: J. Am. Chem. Soc. 107, 4088 (1985)
515. Ley, S. V., Murray, P. J.: J. Chem. Soc., Chem. Commun. 1252 (1982)
516. Ley, S. V., Murray, P. J., Palmer, B. D.: Tetrahedron 41, 4765 (1985)
517. Disanayaka, B. W., Weedon, A. C.: J. Chem. Soc., Chem. Commun. 1282 (1985)
518. McMurry, J. E.: Acc. Chem. Res. 16, 405 (1983)
519. Hewson, A. T., MacPherson, D. T.: J. Chem. Soc. Perkin Trans. I, 2625 (1985)
520. Ayanoglu, E., Grebreyesus, T., Beecham, C. M., Djerassi, C.: Tetrahedron Lett. 1671 (1978)
521. Kaisin, M., Sheikh, Y. M., Durham, L. J., Djerassi, C., Tursch, B., Daloze, D., Braekman, J. C., Losman, D., Karlsson, R.: ibid. 2239 (1974)

522. Sheikh, Y. M., Singy, G., Kaisin, M., Eggert, H., Djerassi, C., Tursch, B., Daloze, D., Braekman, J. C.: Tetrahedron 32, 1171 (1976)
523. Sheikh, Y. M., Djerassi, C., Braekman, J. C., Daloze, D., Kaisin, M., Tursch, B., Karlsson, R.: ibid. 33, 2115 (1977)
524. Karlson, R.: Acta Crystallogr. Sect. B. 32, 2609 (1976)
525. Kaisin, M., Tursch, B., Declercq, J. P., Germain, G., van Meerssche, M.: Bull. Soc. Chim. Belg. 88, 253 (1979)
526. Cierzsko, L. S.: Trans. N. Y. Acad. Sci. 24, 502 (1962)
527. Burkolder, P. R., Burkolder, L. M.: Science 127, 1174 (1958)
528. Cierzsko, L. S., Karns, T. K. B.: In Biology and Geology of Coral Reefs, Vol II (eds. Jones, A., Endean, R.) Academic Press, New York, Chapter 6 (1972)
529. Huguet, J., Karpf, M., Dreiding, A. S.: Helv. Chim. Acta 65, 2413 (1982)
530. Little, R. D., Carroll, G. L.: Tetrahedron Lett. 4389 (1981)
531. Stevens, K. E., Paquette, L. A.: ibid. 4393 (1981)
532. Little, R. D., Carroll, G. L., Peterson, J. L.: J. Am. Chem. Soc. 105, 928 (1983)
533. Pattenden, G., Birch, A. M.: J. Chem. Soc. Perkin Trans. I, 1913 (1983)
534. Birch, A. M., Pattenden, G.: J. Chem. Soc., Chem. Commun. 1195 (1980)
535. Birch, A. M., Pattenden, G.: Tetrahedron Lett. 991 (1982)
536. Mehta, G., Reddy, D. S., Murty, A. N.: J. Chem. Soc., Chem. Commun. 824 (1983)
537. Paquette, L. A., Stevens, K. E.: Can. J. Chem. 62, 2415 (1984)
538. Crisp, G. T., Scott, W. J., Stille, J. K.: J. Am. Chem. Soc. 106, 7500 (1984)
539. Santelli-Rouvier, C., Santelli, M.: Synthesis 429 (1983)
540. Liu, H. J., Kulkarni, M. G.: Tetrahedron Lett. 4847 (1985)
541. Curran, D. P., Chen, M.-H.: ibid. 4991 (1985)
542. Piers, E., Karunaratne, V.: J. Chem. Soc., Chem. Commun. 935 (1983)
543. Piers, E., Karunaratne, V.: Can. J. Chem. 62, 629 (1984)
544. Stille, J. R., Grubbs, R. H.: J. Am. Chem. Soc. 108, 855 (1986)
545. Tebbe, F. N., Parshall, G. W., Reddy, G. S.: ibid. 100, 3611 (1978)
546. Shibasaki, M., Mase, T., Ikegami, S.: Chem. Lett. 1737 (1983)
547. Shibasaki, M., Mase, T., Ikegami, S.: J. Am. Chem. Soc. 108, 2090 (1986)
548. Takeuchi, T., Iinuma, H., Iwanaga, J., Takahashi, S., Takita, T., Umezawa, H.: J. Antibiot. 22, 215 (1969);
Takeuchi, T., Iinuma, H., Takahashi, S., Umezawa, H.: ibid. 24, 631 (1971)
549. Schuda, P. F., Ammon, H. L., Heimann, M. R., Bhattacharjee, S.: J. Org. Chem. 47, 3434 (1982);
Schuda, P. F., Heimann, M. R.: Tetrahedron Lett. 4267 (1983);
Schuda, P. F., Heimann, M. R.: Tetrahedron 40, 2365 (1984);
Schuda, P. F., Heimann, M. R.: Ibid. 40, 4159 (1984)
550. Lansbury, P. T., Wang, N. Y., Rhodes, J. E.: Tetrahedron Lett. 1829 (1971);
Lansbury, P. T., Wang, N. Y., Rhodes, J. E.: ibid. 2053 (1972)
551. Tatsuka, K., Akimoto, K., Kinoshita, M.: J. Antibiot. 33, 100 (1980);
Taksuka, K., Akimoto, K., Kinoshita, M.: Tetrahedron 37, 4365 (1981)
552. Danishefsky, S., Zamboni, R., Kahn, M., Etheredge, S. J.: J. Am. Chem. Soc. 102, 2097 (1980);
Danishefsky, S., Zamboni, R., Kahn, M., Etheredge, S. J.: ibid. 103, 3460 (1981);
Danishefsky, S., Zamboni, R.: Tetrahedron Lett. 3439 (1980)
553. Shibasaki, M., Iseki, K., Ikegami, S.: Synth. Commun. 10, 545 (1980);
Shibasaki, M., Iseki, K., Ikegami, S.: Tetrahedron Lett. 3587 (1980);
Iseki, K., Yamazaki, M., Shibasaki, M., Ikegami, S.: Tetrahedron 37, 4411 (1981)

554. Exon, C., Magnus, P.: J. Am. Chem. Soc. *105*, 2478 (1983);
     Magnus, P., Exon, C., Albaugh-Robertson, P.: Tetrahedron *41*, 5861 (1985)
555. Trost, M., Curran, D.: J. Am. Chem. Soc. *103*, 7380 (1981)
556. Koreeda, M., Mislankar, S. G.: ibid. *105*, 7203 (1983)
557. Ito, T., Tomiyoshi, N., Nakamura, K., Azuma, S., Izawa, M., Maruyama, F., Yanagiya, M., Shirahama, H., Matsumoto, T.: Tetrahedron Lett. 1721 (1982)
558. Wender, P. A., Howbert, J. J.: ibid. 5325 (1983)
559. Ito, T., Tomiyoshi, N., Nakamura, K., Azuma, S., Izawa, M., Maruyama, F., Yanagiya, M., Shirahama, H., Matsumoto, T.: Tetrahedron *40*, 241 (1984)
560. Demuth, M., Ritterskamp, P., Schaffner, K.: Helv. Chim. Acta *67*, 2023 (1984)
561. Van Hijfte, L., Little, R. D.: J. Org. Chem. *50*, 3942 (1985)
562. Heatley, N. G., Jennings, M. A., Florey, H. W.: Br. J. Exp. Pathol. *28*, 35 (1947)
563. Feline, T. C., Mellows, G., Jones, R. B., Phillips, L.: J. Chem. Soc., Chem. Commun. 63 (1974)
564. Yamazaki, M., Shibasaki, M., Ikegami, S.: Chem. Lett. 1245 (1981)
565. Shibasaki, M., Yamazaki, M., Iseki, K., Ikegami, S.: Tetrahedron Lett. 5311 (1982)
566. Yamazaki, M., Shibasaki, M., Ikegami, S.: J. Org. Chem. *48*, 4402 (1983)
567. Greene, A. E., Luche, M.-J., Deprés, J.-P.: J. Am. Chem. Soc. *105*, 2435 (1983)
568. Greene, A. E., Luche, M.-J., Serra, A. A.: J. Org. Chem. *50*, 3957 (1985)
569. Hashimoto, H., Tsuzuki, K., Sakan, F., Shirahama, H., Matsumoto, T.: Tetrahedron Lett. 3745 (1974)
570. Zalkow, L. H., Harris, R. N., Van Derver, D., Bertrand, J. A.: J. Chem. Soc., Chem. Commun. 456 (1977)
571. Bohlmann, F., Le Van, N., Pickardt, J.: Chem. Ber. *110*, 3777 (1977)
572. Zalkow, L. H., Harris, R. N., Burke, N. I.: J. Nat. Prod. *42*, 96 (1979)
573. Wenkert, E., Arrhenius, S. T.: J. Am. Chem. Soc. *105*, 2030 (1983)
574. Short, R. P., Revol, J.-M., Ranu, B. C., Hudlicky, T.: J. Org. Chem. *48*, 4453 (1983)
575. Ranu, B. C., Kavka, M., Higgs, L. A., Hudlicky, T.: Tetrahedron Lett. 2447 (1984)
576. Kwart, L. D., Tiedje, M., Frazier, J. O., Hudlicky, T.: Synth. Commun. *16*, 393 (1986)
577. Chatterjee, S.: J. Chem. Soc., Chem. Commun. 620 (1979)
578. Tobe, Y., Yamashita, T., Kakiuchi, K., Odaira, Y.: ibid. 898 (1985)
579. Bohlmann, F., Jakupovic, J.: Phytochemistry *19*, 259 (1980)
580. Leone-Bay, A., Paquette, L. A.: J. Org. Chem. *47*, 4173 (1983)
581. Tsunoda, T., Kodama, M., Ito, S.: Tetrahedron Lett. 83 (1983)
582. Sternbach, D. D., Hughes, J. W., Burdi, D. F., Banks, B. A.: J. Am. Chem. Soc. *107*, 2149 (1985)
583. Wender, P. A., Ternansky, R. J.: Tetrahedron Lett. 2625 (1985)
584. Seto, H., Yonehara, H.: J. Antibiot. *33*, 92 (1980)
585. Cane, D. E., Abell, C., Tillman, A. M.: Bioorg. Chem. *12*, 312 (1984)
586. Misumi, S., Ohtsuka, T., Ohfune, Y., Sugita, K., Shirahama, H., Matsumoto, T.: Tetrahedron Lett. 31 (1979)
587. Paquette, L. A., Annis, G. D.: J. Am. Chem. Soc. *104*, 4504 (1982)
588. Piers, E., Karunaratne, V.: J. Chem. Soc., Chem. Commun. 959 (1984)
589. Pattenden, G., Teague, S. J.: Tetrahedron Lett. 3021 (1984)
590. Baker, R., Keen, R. B.: J. Organomet. Chem. *285*, 419 (1985)
591. Mehta, G., Srinivas Rao, K.: J. Chem. Soc., Chem. Commun. 1464 (1985)
592. Seto, H., Sasaki, T., Uzawa, J., Takeuchi, S., Yonehara, H.: Tetrahedron Lett. 4411 (1978)

593. Sakai, K., Ohtsuka, T., Misumi, S., Shirahama, H., Matsumoto, T.: Chem. Lett. 355 (1981)
594. Cane, D. E., Rossi, T., Pachlatko, J. P.: Tetrahedron Lett. 3639 (1979);
     Cane, D. E., Rossi, T., Tillman, A. M., Pachlatko, J. P.: J. Am. Chem. Soc. *103*, 1838 (1981)
595. Takahashi, S., Takeuchi, M., Arai, M., Seto, H., Otake, N.: J. Antibiot. *36*, 226 (1983)
596. Crimmins, M. T., DeLoach, J. A.: J. Am. Chem. Soc. *108*, 800 (1986)
597. Crimmins, M. T., DeLoach, J. A.: J. Org. Chem. *49*, 2077 (1984)
598. Bohlmann, F., Zdero, C.: Phytochemistry *18*, 1747 (1979)
599. Paquette, L. A., Galemmo, R. A., Springer, J. P.: J. Am. Chem. Soc. *105*, 6975 (1983)
600. Tsunoda, T., Kabasawa, Y., Ito, S.: Tetrahedron Lett. 773 (1984)
601. Paquette, L. A., Galemmo, R. A., Caille, J.-C., Valpey, R. S.: J. Org. Chem. *51*, 686 (1986)
602. Tori, M., Matsuda, R., Asakawa, Y.: Bull. Chem. Soc. Jpn. *58*, 2523 (1985)
603. Bohlmann, F., Jakupovic, J.: Phytochemistry *19*, 259 (1980)
604. Bohlmann, F., Suding, H., Cuatrecasos, J., Robinson, H., King, R. M.: ibid. *19*, 2399 (1980)
605. Paquette, L. A., Roberts, R. A., Drtina, G. J.: J. Am. Chem. Soc. *106*, 6690 (1984)
606. Marx, J. N., Naman, L. R.: J. Org. Chem. *40*, 1602 (1975)
607. Marfat, A., Helquist, P.: Tetrahedron Lett. 4217 (1978)
608. Stowell, J. C.: J. Org. Chem. *41*, 560 (1976);
     Stowell, J. C., Keith, D.: Synthesis 132 (1979)
609. Wender, P. A., Singh, S. K.: Tetrahedron Lett. 5987 (1985)
610. Curran, D. P., Kuo, S.-K.: J. Am. Chem. Soc. *108*, 1106 (1986)
611. Rao, P. S., Sarma, K. G., Seshardri, T. R.: Curr. Sci. *34*, 9 (1965);
     Rao, P. S., Sarma, K. G., Seshardri, T. R.: ibid. *35*, 147 (1966)
612. Kaneda, M., Takahashi, R., Iitaka, Y., Shibata, S.: Tetrahedron Lett. 4609 (1972)
613. Corey, E. J., Desai, M. C., Engler, T. A.: J. Am. Chem. Soc. *107*, 4339 (1985)
614. Zalkow, L. H., Harris, R. N., Van Derveer, D.: J. Chem. Soc., Chem. Commun. 420 (1978)
615. Smith, A. B., Jerris, P. J.: J. Am. Chem. Soc. *103*, 194 (1981)
616. Karpf, M., Dreiding, A. S.: Tetrahedron Lett. 4569 (1980)
617. Schostarez, H., Paquette, L. A.: J. Am. Chem. Soc. *103*, 722 (1981)
618. Oppolzer, W., Marazza, F.: Helv. Chim. Acta *64*, 1575 (1981);
     Oppolzer, W., Bättig, K.: ibid. *64*, 2489 (1981)
619. Wender, P. A., Dreyer, G. B.: J. Am. Chem. Soc. *104*, 5805 (1982)
620. Wrobel, J., Takahashi, K., Honkan, V., Lanneye, G., Cook, J. M., Bertz, S. H.: J. Org. Chem. *48*, 139 (1983)
621. Tobe, Y., Yamashita, S., Yamashita, T., Katiuchi, K., Odaira, Y.: J. Chem. Soc., Chem. Commun. 1259 (1984)
622. Wilkening, D., Mundy, B. P.: Tetrahedron Lett. 4619 (1984)
623. Mundy, B. P., Wilkening, D., Lipkowitz, K. B.: J. Org. Chem. *50*, 5727 (1985)
624. Mehta, G., Subrahmanyan, D.: J. Chem. Soc., Chem. Commun. 768 (1985)

Subject Index